EXPLORING C FOR MICROCONTROLLERS

Exploring C for Microcontrollers

A Hands on Approach

JIVAN S. PARAB
Goa University
India

VINOD G. SHELAKE
Shivaji University
Kolhapur, India

RAJANISH K. KAMAT
Shivaji University
Kolhapur, India

and

GOURISH M. NAIK
Goa University
India

A C.I.P. Catalogue record for this book is available from the Library of Congress.

ISBN 978-1-4020-6066-3 (HB)
ISBN 978-1-4020-6067-0 (e-book)

Published by Springer,
P.O. Box 17, 3300 AA Dordrecht, The Netherlands.

www.springer.com

Printed on acid-free paper

Contents

Foreword

If we accept the premise that an embedded engineer is made rather than born, then how does one go about making a good one? The authors of this book *Exploring C for Microcontrollers: A Hands-on Approach* are certainly "good ones". Not only do they explore some of the influences that shaped themselves but they also try to shape "would-be" embedded engineers. Research and developmental activities in embedded systems has grown in a significant proportion in the recent past. Embedded software design is not new to the world, but with the changing time, it has gained considerable momentum in the recent past, and many young engineers are strongly inclined to pursue their future in this field. The book is mainly targeted to these engineers who would like to understand in great depth the synergetic combination of hardware and software.

The book is divided into eight chapters. Chapter 1 introduces a brief background about micro-controllers and explains how they are embedded into products commercially available in the market to emphasize the importance of these in the daily life of mankind. It also gives an insight into the architectural details and embedded system concepts for students' projects to motivate them into this exciting field. The rest of the book concentrates on software development. The integrated development environment (IDE) is introduced in Chapter 2. Again the screen shots and step-by-step procedure will certainly make the students and engineers fully understand the development process. Chapter 3 differentiates the embedded C paradigm from the conventional ANSI C. Again the authors explain how to successfully overcome the memory and time constraints while developing an embedded C program. Chapter 4 gives an overview of program development for on-chip resources for MCS51 family of microcontrollers. Chapters 5–8 are devoted to live case studies.

The book has come out with an elegant presentation to aspiring students and engineers from the teaching experience and technical knowledge the authors have put over a long time in this field. I strongly recommend this book for intermediate programmers, electronics, electrical, instrumentation engineers or any individual who is strongly inclined to take up his or her career in embedded C programming. I am sure the reader will experience learning embedded programming by example and learning by doing. Last but not the least, this book will certainly be a value addition to the world of embedded programming.

Dr. A. Senthil Kumar
Head
Data Quality Evaluation
National Remote Sensing Agency
Department of Space
Government of India

Dr. Senthil Kumar is Head of DQE and PQCD sections of National Remote Sensing Agency (NRSA) an autonomous operational center under Department of Space (DOS), Government of India. This is the nodal agency in the country for receiving, processing, and distributing the satellite and aerial remote sensing data and products. NRSA is also responsible for providing end-to-end solutions for utilization of data for geospatial applications and information services. NRSA has a huge archive of remote sensing data acquired through Indian and foreign satellites and also has the capability to acquire data pertaining to any part of the globe on demand. It is one of the important centers for promotion of remote sensing and geographic information system technologies in India. NRSA has set up satellite data processing facilities starting from data reception to utilization at various centers within India and across the globe.

Preface

The past few decades have witnessed evolution of microcontrollers. They have revitalized a number of products or equipment in almost all fields including telecommunications, medical, industrial, and consumer products. These embedded microcontroller systems now resides at the heart of modern life with a variety of applications in fields like consumer electronics, automotive systems, domestic, and even in aerospace products. Embedding a microcontroller in an electronics instrument or product requires a specialized design skill which requires a synergy of hardware and software skills.

In our day-to-day life we come across a number of embedded products. When we switch on the washing machine or send an SMS on a cell phone one cannot prevent without thinking the mechanism and the co-working of hardware and software in the background. The market for such smart embedded products is occupying newer and newer applications seemingly impossible few years back. Last year the IDC, a premier global market intelligence firm, revealed that the embedded industry product development was expected to be as high as $75 billion. This entails the industry requirement of trained human resource with mixed skill set both in hardware and software. Unfortunately the synergetic demand of hardware and software or sometimes even referred to as firmware competency has led to a supply–demand gap of HR in this field. This gap expressed in numerical figures led to requirement of around 150,000 embedded engineers to serve the global embedded industry. This book is ideal for all those who would like to pursue their career in the exciting world of microcontroller-based embedded systems. The approach is pedagogical; first the hardware module is presented and then the associated software code in Keil C.

The hardware designed is useful for engineering graduates and practicing professionals with the required knowledge and practical hands on skills to design with embedded systems. However, the prerequisite for the book is background of theoretical aspects of architecture of microcontrollers especially the MCS-51 family. The book starts with initial experiments, which provide familiarization with the capabilities and the limitations of the basic 8051 microcontroller using a simulator. Once the reader is comfortable with these primitive programs which covers almost all the on-chip resources, he or she can switch to more advanced ones.

The Scope of the Book

We now review the topics covered in sequence, chapter by chapter.

Chapter 1 provides an overview of microcontrollers and their applications in different domains. The architectural trends and the growth economics emphasizes the importance of the subject. The photograph of the setup and the hints toward project execution will definitely boost the confidence of the novice to kick-start the project with minimal resources.

Chapter 2 is devoted to the IDE for the MCS-51 family. The simulation and single stepping as described in this chapter will solve all the project intricacies of the readers. Chapter 3 illustrates the basic difference in traditional C programming and embedded C. Chapter 4 deals with the programming of on-chip resources of MCS-51 family microcontrollers in C. The theoretical details of these on-chip resources such as ports, timers, etc., are completely eliminated. As the book aims at hands-on approach, the programs for the on-chip resources have been developed and their execution is illustrated in the Keil simulation environment.

The last four chapters, i.e., 5–8 deal with various project case studies. Several case studies in various application domains such as lighting, measurement and control, security, and domestic applications are developed from scratch. The hardware and software developed in the form of case studies also caters to a set of mini projects, which are discussed in detail from the design phase to the actual implementation on a target system. There are 17 case studies given in this book on various systems that you may encounter in day-to-day life. Overall the hardware and software developed in this book can be reused for any embedded system project and is expected to act as a rapid prototyping unit for the embedded systems industry.

Reasons for Proposing this Book

The market is flooded with a number of good books on embedded systems designed especially with the most popular MCS-51 family. These books are traditional in nature, i.e., they start with the routine architectural features of 8051, description of registers, ports, interrupts, etc. Most of these are already covered in the device data sheet and application notes. In the present book all such routine features are skipped. The focus is on programming microcontrollers to be specific MCS-51 family in 'C' using Keil IDE. The book presents 20 live case studies apart from the many basic programs organized around every on-chip resource like port, time/counter, interrupt, serial I/O, etc. Rather than introducing the underpinning theory or reproducing lengthy data sheets, our approach is "learning-through-doing" and one that appeals to busy electronics designers. The 'C' codes given are well supported by easy-to-understand comments wherever required. Mastering the basic modules and hands-on working with the projects will enable the reader to grasp the basic building blocks for most 8051 programs. Whether you are a student using the MCS-51 family of microcontrollers for your project work or an embedded systems programmer, this book will give you a kick start in using and understanding the most popular microcontroller.

Authors through their interaction with the undergraduate and post-graduate students as well as industry professionals have found that such a book is the need of the microcontroller community interested in C programming. The book will bridge the gap between the microcontroller hardware experts and the C programmers.

Major Features

The objective of this book is to introduce the readers to the design and implementation of an embedded system. It covers the unique requirements and limitations of operating in an embedded environment. It also covers microcontrollers as the most widespread example of embedded systems. In particular, it focuses on the MCS-51 family of microcontrollers, their programming in C language, and interfacing techniques.

Special emphasis is to provide hands-on experience for the readers using a hardware and interfacing modules described in this book. The aim is to empower the reader to actually solve his or her problem with a practical hands-on pedagogy through the hardware and software presented in this book. The principle of "Design Reuse" is explained effectively.

Further, the readers will also learn how to follow the sequence of data flow through the microcontroller when a program is executed. Additionally, the readers will learn the operation of the microcontroller's I/O functions and the external devices driven by the microcontroller. Hardware and software design issues are discussed for specific systems implemented using MCS-51 as the embedded microcontroller.

Acknowledgments

We would like to take the opportunity to thank all those who have contributed or helped in some way in the preparation of this text. Particular thanks must go to our heads of the institutions – Professor M.M. Salunkhe, Vice Chancellor, Shivaji University, Kolhapur, India, and Professor P.S. Zacharias, Vice Chancellor, Goa University, Goa, India – for the encouragement and support. We would also like to thank Mr. P. Venugopal, Director, Software Technology Parks of India, Maharashtra Region, for his support. Dr. Kamat and Dr. Naik would like to thank their respective wives for their understanding and patience shown when the preparation of the book took time which could have been spent with the family. Our thanks then to Kamat's wife Rucha and Naik's wife Deepa.

Additionally, Mr. Shelake and Mr. Parab would like to express gratitude to their parents for their encouragement and support over the years. Kamat would like to dedicate his contribution to this book to the memory of the late Professor G.G. Tengshe and the late Dr. V. Rao Indolkar, ACD Machine Control Tools Ltd., Mumbai.

Dr. Kamat would also like to thank to his teacher in this field Mr. S. Ramgopal, Indian Institute of Science, Bangalore, and Dr. Senthil Kumar, Dr. Raghurama, Deputy Director (Academic) of BITS Pilani as well as Mr. K.S. Deorukhkar for help in reviewing and critical suggestions. The past batches of M.Sc. Electronics students of both Shivaji University and Goa University especially Mr. Roy, Mr. Rupesh from Satyam Computers must be thanked for generation of problems for programs developed in this book.

- Jivan S. Parab
- Vinod G. Shelake
- Dr. Rajanish K. Kamat
- Dr. Gourish M. Naik

Chapter 1

Microcontrollers: Yesterday, Today, and Tomorrow

1.1 Defining Microcontrollers

It is said that from the definition everything true about the concept follows. Therefore, at the outset let us take a brief review of how the all-pervasive microcontroller has been defined by various technical sources.

A microcontroller (or MCU) is a computer-on-a-chip used to control electronic devices. It is a type of microprocessor emphasizing self-sufficiency and cost-effectiveness, in contrast to a general-purpose microprocessor (the kind used in a PC). A typical microcontroller contains all the memory and interfaces needed for a simple application, whereas a general purpose microprocessor requires additional chips to provide these functions....(Wikipedia [1])

A highly integrated chip that contains all the components comprising a controller. Typically this includes a CPU, RAM, some form of ROM, I/O ports, and timers. Unlike a general-purpose computer, which also includes all of these components, a microcontroller is designed for a very specific task – to control a particular system. As a result, the parts can be simplified and reduced, which cuts down on production costs. ... (Webopedia [2])

A {microprocessor} on a single {integrated circuit} intended to operate as an {embedded} system. As well as a {CPU}, a microcontroller typically includes small amounts of {RAM} and {PROM} and timers and I/O ports....(Define That [3])

A single chip that contains the processor (the CPU), non-volatile memory for the program (ROM or flash), volatile memory for input and output (RAM), a clock and an I/O control unit. ... (PC Magazine [4])

A microprocessor on a single integrated circuit intended to operate as an embedded system. As well as a CPU, a microcontroller typically includes small amounts of RAM and PROM and timers and I/O ports. ... (FOLDOC [5])

J.S. Parab et al. (eds.), Exploring C for Microcontrollers, 1–18.

The definitions given by various sources describe microcontroller as an integrated circuit (IC) with processor as well as peripherals on chip. But the crux of the matter is the widespread uses of microcontrollers in electronic systems. They are hidden inside a surprising number of products such as microwave oven, TV, VCR, digital camera, cell phone, Camcorders, cars, printers, keyboards, to name a few.

The last three decades and especially the past few years have witnessed the tremendous growth in the top-of-the-line processors used for personal computing and digital signal processor (DSP) applications. Today, the use of microcontrollers in embedded systems outnumbers the use of processors in both the PC and the workstation market. It is estimated that over 50% of the vast majority of the computer chips sold are microcontrollers. The main reasons behind their huge success are powerful yet cleverly chosen customizable electronics, ease in programming, and low cost. These days microcontrollers find increasing application even in areas where they could not be used previously. With the decreasing costs and footprints, it is now possible to have small-sized and cost-effective microcontroller units (MCUs) for new applications. The microcontrollers today are small enough to penetrate into the traditional market for 4-bit applications like TV remote controls, toys, and games. For the simplest of all control applications they can offer high value smart switch functionality for applications that previously used electromechanical devices. Also the customers now can add intelligence to their end products for low cost as per the microcontroller market report by Frost & Sullivan research service [6].

1.2 Eagle's View: Microcontrollers and Other Competing Devices

Generally the technical fraternity try to compare the various devices like microprocessors, PCs, microcontrollers, DSPs, and reconfigurable devices like FPGAs and CPLDs. An interesting point to note is that embedded systems are made using all the above-mentioned devices except PC owing to its general purpose architecture. As programmability is the common feature of all these devices, they have their firm footing in different application domains. On one side of the spectrum, microcontroller-based embedded system design emphasizes on task-specific dedicated applications at low power, low cost, high throughput, and highest reliability. On the other extreme of the spectrum, FPGA-based embedded systems dominate their custom computing architectures. Unlike microcontrollers, these systems can be reconfigured on the fly to suit the application with higher computational density and throughput. With the proliferation of density, FPGA-based embedded systems offer higher performance with the only challenging issue of memory required to store their configurations.

The technical community also tends to associate various characteristics of embedded systems with microprocessors and microcontrollers. The microprocessors are typically found to dominate the desktop arena focusing on more and more bit processing capability, with other features such as fairly good amount of cache with memory management and protection schemes supported by the operating system. Although microcontrollers share flexibility aspect of microprocessors through programming, 8-bit versions are more in use (although 16- and 32-bit exist) with RAM and ROM (instead of cache) added with many on chip peripherals like timer/counter, decoder, communication interface, etc. In the literature many embedded systems products have been reported as microprocessors. On the other side of the processor spectrum, a DSP possesses special architecture that provides ultra-fast instruction sequences, such as shift and add, multiply and add, which are commonly used in math-intensive signal processing applications. The common attributes associated with the DSPs are multiply-accumulate (MAC) operations, deep pipelining, DMA processing, saturation arithmetic, separate program and data memories, and floating point capabilities required most of the time. However, the line of differentiation between all these devices is getting blurred at a rapid pace. With the introduction of fuzzy logic, artificial intelligence and networked communication capabilities in the consumer products like refrigerators, mobile phones, and cars, convergence of the architectures of most of the above-mentioned programmable devices is witnessed by the industry. Today's ideal microcontroller is expected to offer plenty of MIPS, run the DSP programs with the same speed of the DSP processor, integrate all its peripherals and support flash, communicate with the world with I2C or CAN protocols, withstand extremes of environment in a car engine, and cost but a few cents.

1.3 Vignettes: Microcontrollers

It is interesting to note that the development of microprocessors seems to be an accident out of the microcontroller synthesis. In 1969, Busicom, a Japanese company, approached Intel to convert their special purpose ROM and shift register–based calculator cores into a specialized application specific processor. The objective was the development of microcontrollers rather than a general purpose of CPU chips for keyboard scanning, display control, printer control, and other functions for a Busicom-manufactured calculator. However, the Intel engineers opted for a more flexible programmable microcomputer approach rather than the random logic chip-set originally envisioned by Busicom. The four designs [7] proposed by Federico Faggin, Ted Hoff, and Stan Mazor from Intel were a 2048-bit ROM with a 4-bit programmable input/output port (4001); a 4-registers × 20-locations × 4-bit RAM data memory with a 4-bit output port (4002); an input/output (I/O) expansion chip, consisting of a static shift register with serial input and serial/parallel output (4003); and the 4-bit CPU chip (4004). The 4001, 4002, and 4003 are very close to microcontroller kind of architecture rather than microprocessor. However, the Intel 4004, which was supposed to be the brains of a calculator, turned out to be a general-purpose microprocessor as powerful as ENIAC. The scientific papers and literature published around 1971 reveal that the MP944 digital processor used for the F-14 Tomcat aircraft of the US Navy qualifies as the "first microprocessor". Although interesting, it was not a single-chip processor, and was not general purpose – it was more like a set of parallel building blocks you could use to make a special purpose DSP form [8]. It indicates that today's industry theme of converging DSP-microcontroller architectures was started in 1971.

The other companies were also catching up at the same time. The first official claim of filing the patent goes to Texas Instruments under the trade name TMS1000 way back in 1974. This was the first microcontroller which included a 4-bit accumulator, 4-bit Y register and 2- or 3-bit X register, which combined to create a 6- or 7-bit index register for the 64 or 128 nibbles of on-chip RAM. A 1-bit status register was used for various purposes in different contexts. This microcontroller served as the brain of the Texas Instrument's educational toy named "Spark and Spell" shown in the movie *ET: The Extraterrestrial.* In 1976, Intel introduced the first 8-bit microcontroller family MCS-48 which was so popular that they could ship 251,000 units in that year. After four years of continuous research, the MCS-48 family was upgraded to 8051, an 8-bit microcontroller with on-board EPROM memory. Intel shipped 22 million pieces in 1980. The market requirement was so much that the total units sold in three years later were 91 million. The year 2005 is a special one for the microcontroller 8051. It has celebrated its 25th anniversary. But, also in 2005, Intel notified they would discontinue all automotive versions of their microcontrollers, including 8051. Car engine control units were once perhaps the most prominent application for 8051s. This means only one thing, Intel gives up the microcontrollers for good. This is confirmed by product change notification published in early 2006, announcing that Intel drops its whole microcontroller business [9].

1.4 Microcontroller Applications

The microcontroller applications are mainly categorized into the following types (see Figure 1.1):

- Audio
- Automotive
- Communication/wired
- Computers and peripherals
- Consumer
- Industrial
- Imaging and video
- Medical
- Military/aerospace
- Mobile/wireless
- Motor control
- Security
- General Purpose
- Miscellaneous

Automobile industry is the main driving force in propelling the growth of microcontrollers. It is estimated that the microcontrollers constitute 33% of the semiconductors used in a vehicle [10]. Requirements of the automobile sector has forced the microcontroller manufacturers to come out with the new bus protocols such as control area network (CAN) and local interconnect network (LIN). Microcontrollers of all bit cores are used in vehicles according to the Frost & Sullivan Industry report. The 8- and 16-bit microcontrollers are used for low-end applications and lower-cost vehicles while the 32-bit microcontrollers are used for high-end application and high-end vehicles. It is estimated that currently 30–40 microcontrollers are used in low-end vehicles and about 70 microcontrollers are used in high-end vehicles. These requirements are continuously increasing and it is highly likely that the count of microcontrollers in vehicles will further increase in the future, quotes World Microcontrollers Market Report by Frost & Sullivan [10].

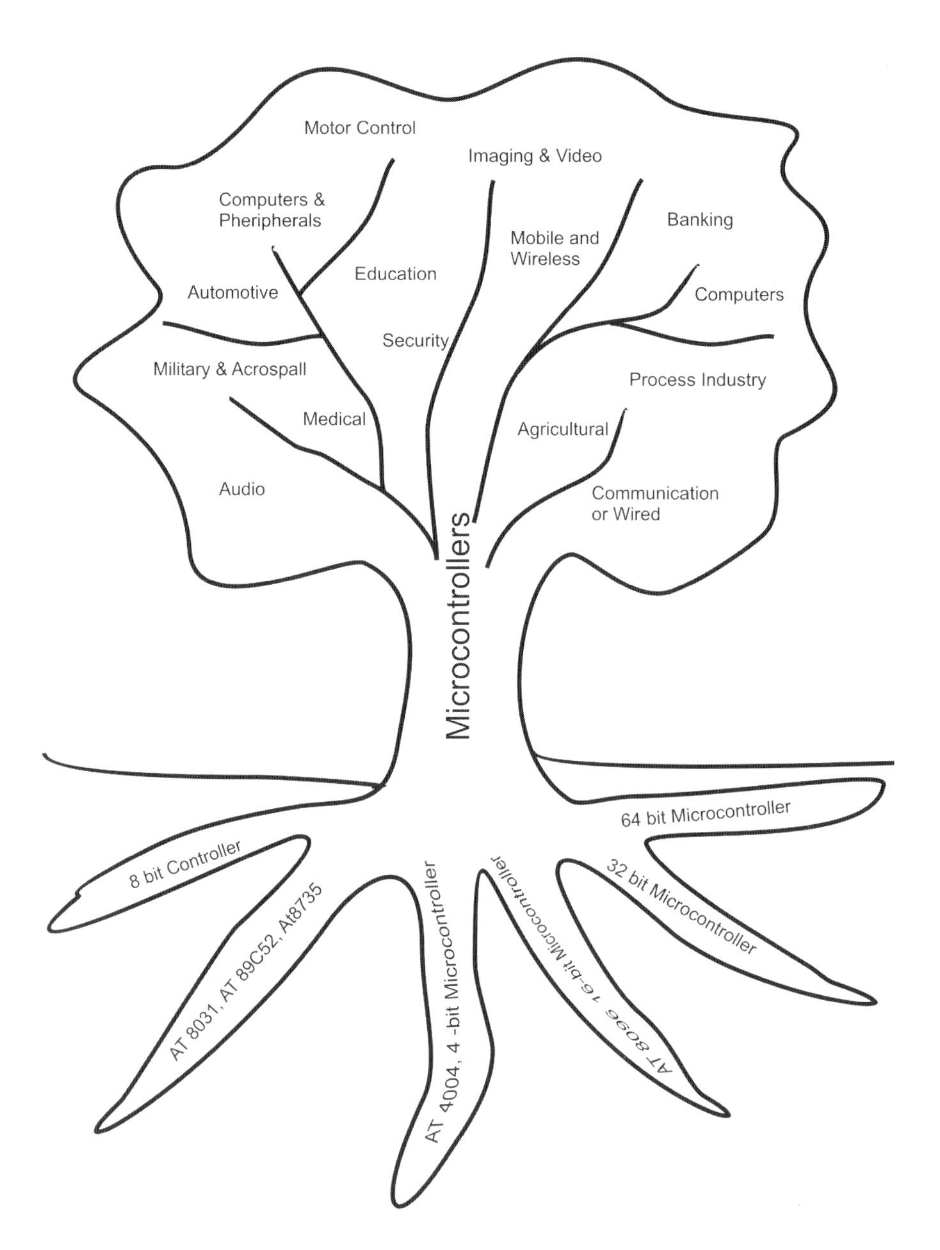

Figure 1.1 Microcontroller application tree

Embedding microcontrollers in the product offers some unique advantages. For an example, in the latest technology washing machines, a transmission is no longer required because a lower-cost AC induction or reluctance motor controlled by sophisticated microcontroller-based electronics can provide all the normal machine cycles [11]. Additionally, the electronically controlled induction or reluctance motor provides a more effective and gentler agitation (wash) cycle that allows the drum containing the clothes to be rotated first in one direction, then stopped, and rotated in the opposite direction without requiring any additional mechanical device. This forward/reverse agitation cycle provides a more effective means of cleaning your clothes without damaging the fibers used to make them.

It is also observed that the induction of microcontrollers in a product has increased the market demand of the product itself. One such example is NEC Electronics' 8-bit microcontroller [12], which is employed in over half of the digital cameras produced throughout the world, thus making it a hit product – albeit one that plays its role behind the scenes. In 2003, the shipment volume of 8-bit microcontrollers for digital camera use achieved monthly shipments of two million. Currently, the most commonly used microcontroller for digital cameras is the μPD78003×. The industry's top-level, low-voltage (1.8 V) A/D converter is built into this compact 64-pin QFP package with an edge measuring a mere 10 × 10 mm.

1.5 Growth Economics

Many industry analysts have reported a bright growth economics for microcontroller in the near future. Bourne Research reports that MEMS-based motion, pressure, and acoustic sensors (as well as other next-generation sensing technologies) are not only finding their way into classic consumer products like toys, housewares, sporting goods, and clothing, but applications as far-reaching as flooring materials, medical diagnostics, retail fixtures, and vending machines. System complexity is just as varied, but Bourne Research notes that, more often than not, these products will utilize multiple sensors and microcontrollers [13]. Another interesting report by the Semiconductor Industry Association (SIA) reveals that the microcontroller sales are projected to grow by 1.9% to $12.3 billion in 2006 and to $15.4 billion in 2009, a compound annual growth rate (CAGR) of 6.3% [14]. The positive growth economics of microcontroller chips also ushers the associated growth of software such as simulators, cross compilers, and assemblers. The market is witnessing many novel supporting software packages with simple user interface with convenient compiling and debugging tools.

1.6 The Major Players in the Microcontroller Chip Market

Table 1.1 Major players in the microcontroller market. (From 2006 EDN Directory of Microprocessors and Microcontrollers)

Name of the company	*Product range*	*Application domain*
Actel	Core8051 8-bit-microcontroller IP	Consumer, communications, automotive, industrial, and military/aerospace applications
Altera	Nios II/f (fast), Nios II/e (economy), Nios II/s (standard), Soft IP cores	Automotive, consumer, industrial, medical
Altium	Royalty-free, 8- and 32-bit, FPGA-based soft processors, and FPGA-independent TSK3000RISC core	Automotive, industrial, communications, and computer-peripheral
Analog devices	ADuC7128 family of precision analog microcontrollers; ADSP-BF561 Blackfin ADSP-BF534 Blackfin, BF539 with CAN	Audio, automotive, communication/wired, consumer, industrial, imaging and video, medical, motor control, security
Applied Micro Circuits Corp	Embedded PowerPC processors PPC405CR PPC405EP, PPC405EZ, PPC405GP, PPC405GPr, nP405H, etc.	Communication/wired, computers and peripherals, consumer, industrial, imaging and video, medical, mobile/wireless, motor control, security
ARC International	Two configurable, 32-bit processor-core families ARC600 and ARC700	Sound advanced subsystems targeting portable media devices and advanced-definition audio applications
ARM	Range of processor cores, including the ARM7, ARM9, ARM10, and ARM11 families and the new Cortex family featuring Thumb-2 technology	Audio, automotive, communication/wired, computers and peripherals, consumer, industrial, imaging and video, medical mobile/wireless, security
Atmel	AVR, AVR32, ARM7, ARM9, 8051, Teak, and Diopsis DSP cores	Consumer, computer/networking, communications, security/smart cards, automotive, industrial, medical, military, and aerospace applications

Cast	8-, 16-, and 32-bit IP cores including 8051 cores with instruction execution one clock per cycle; configurable 8051, 8-bit Z80 and 16-bit 68000 compatible devices	Automotive, communication/ wired, computers and peripherals, consumer industrial, imaging and video, medical military/aerospace, mobile/wireless motor control
Cirrus Logic	ARM9- and ARM7-based embedded processors EP9301, EP9302, EP9307, etc.	Audio and industrial markets
Dallas Semiconductor	Four microcontroller families targeting networked, secure, mixed-signal, and 8051 drop-in designs	Networking applications as they have an optional TCP/IP stack in ROM, a built-in Ethernet MAC (media access controller), CAN (controller area network), and parallel and serial ports
Freescale Semiconductor	Ultralow-end, RS08-based, 8-bit microcontrollers to an advanced 32-bit ColdFire core	Mobile and videoconferencing phones, portable media players, PDAs, and portable GPS applications, networking, home and SOHO (small-office/home-office) networking, automobile
Fujitsu Microelectronics America	32, 16, and 8-bit microcontrollers with onboard-flash, ROM, ADC, DAC, CAN (controller area network), USB, and LCD controllers	Automotive, communications, computer peripheral, industrial, and consumer applications
Infineon Technologies	8-, 16-, and 32-bit bit XC886/888, XC164CM TC116×	Industrial and automotive, industrial motor control, building control for lifts and escalators, intelligent sensors, distributed I/O modules, and industrial automation
Lattice Semiconductor	8-bit microcontrollers, including the 8051 and PIC, through its partners, Cast and Digital Core Design, LatticeMico8 an 8-bit soft microcontroller core	Automotive, communication/wired, computers and peripherals, consumer industrial, medical
Luminary Micro	ARM Cortex-M3-based microcontrollers	Industrial and motor control
Microchip Technology	8- and 16-bit PIC microcontrollers, as well as 16-bit dsPIC DSCs (digital signal controllers)	Motor control and general purpose

Table 1.1 Continued.

Name of the company	*Product range*	*Application domain*
National Semiconductor	COP8 flash microcontrollers featuring an 8-bit core with 32 kbytes of onboard flash	Hardware communications peripherals, and an expandable external bus to target embedded system communications applications, such as automotive telematics, vehicle-network gateways, hands-free car kits, and industrial and medical instrumentation and control
NEC Electronics America	8-bit uPD78F0711 and uPD78F0712 flash-based microcontrollers, 32-bit flash-based V850ES/IE2 series	Consumer appliances, home health care, building management systems and industrial controlled applications

1.7 Architectural Trends

The emerging trends in microcontroller architecture are dictated by the technological needs of the embedded system applications. In general the common road map is characterized by the features like single functional, tightly constrained in terms of cost, power, speed, and footprint as well as continuously reactive in a real-time manner. The microcontroller world is divided over the architecture used. Controllers like 8051, 68HC11 join their microprocessor counterparts (80×86 family) in using Von-Neumann architecture by sharing RAM and program memory with the same bus. This imposes the same bit width for the buses irrespective of their requirement, which is not the case with the PIC processors which uses Harvard architecture.

8-bit Treads on MCU Turf

It is really interesting to note that in this era of 32-bit processors, the 8-bit microcontrollers are flourishing and enjoying a stable future. The reasons behind this are low cost and inexpensive development software. An 8-bit microcontroller like 8051 today enjoys 40% of the market share [15]. It has become so popular that about 40 manufacturers now make it with 800 derivatives used for diverse applications right from simple process control to high-tech telecom. The original MCS-51 family developed by Intel consists of CHMOS processors with features such as 8-bit CPU optimized for event control, Boolean processing, on-chip memory (up to 32 K) and on-chip peripherals (timer/counters, serial ports, PCA, etc.). The other manufacturers like Atmel, Dallas, Philips, Analog Devices, etc., have extensively worked for strengthening the basic core by introducing additional features such as Additional Data Pointers (DPTR), Extended Stack Space, Fast Math Operations and Extended or Reduced Instruction Sets to name a few. Table 1.1 summarizes the features of the popular 8051 derivatives manufactured by various companies. Although industry analysts predicted the saturation of 8051 family and also its death, it turned out to be a rumor. In 2003, Cygnal released world's highest performance 8051 microcontroller [16]. The C8051F120 family devices include 128 Kbyte Flash memory, 8448 byte RAM, and are packaged in either a 100-pin or 64-pin TQFP. On-chip peripherals include a 12-bit 100 ksps ADC, two 12-bit DACs, a 2% internal oscillator, temperature sensor, voltage reference, two UARTs, and many features of a DSP processor.

Low Power Design

The latest 8-bit devices continue to drive up the performance bar with simplicity for usage and ease of programming. Most of these devices are aimed at low power consumption achieved by using sleep modes and the ability to turn certain peripherals on and off. The best example for this is XE88LC02 from XEMICS [17], a recently launched microcontroller, which features programmable gain/offset amplifier followed by high resolution ADC with four differential or seven pseudo differential inputs, with current consumption as low as 2 μA in real-time clock mode and a typical 300 μA/MIPS in sustained computing mode. Recently, Atmel has also strategically evolved their microcontroller architecture by implementing novel power saving techniques. Atmel's picoPower technology uses a variety of techniques that eliminate unnecessary power consumption in power-down modes. These include an ultra-low-power 32 kHz crystal oscillator, automatic disabling and re-enabling of brown-out detection (BOD) circuitry during sleep modes, a power reduction register that completely powers down individual peripherals, and digital input disable registers that reduces the leakage current on digital inputs [18]. PicoPower microcontrollers consume down to 340 μA in active mode, 150 μA in idle mode at 1 MHz, 650 nA in power-save mode and 100 nA in power-down mode.

Convergence of DSP with Micro

The trend of convergence of functionality in microcontroller was started in the 1970s with MP944, which had more of DSP orientation. For a microcontroller engineer to consider a DSP or for a DSP engineer to use a microcontroller, three strict criteria must be met: price/performance, peripheral set, and development tool quality [19]. This trend is now more pronounced as seen in the newly launched microcontroller cores by Motorola. For instance, in its 56F83xx family of microcontrollers, Motorola has introduced a series of devices that it intends to meet uprated processing needs in today's systems – particularly in automotive systems – while keeping 8/16-bit economics and avoiding the shift to 32-bit controllers. The defining features of the device series are the inclusion of some DSP attributes and the use of on-chip flash memory instead of EEPROM. The term "hybrid" describes the devices' combination of microcontroller and DSP functions. These devices are not hybrid in the sense that they combine twin processing cores; rather, they add the facility to efficiently execute certain DSP-like arithmetic functions such as multiply-accumulate operations [20]. These kinds of architectures are in heavy demand owing to such hybrid signal processing requirements in application domains such as speech processing in consumer applications. The key challenges in converging these two diversified architectures are combining the pipelining, register oriented, complex conditional resolving architecture of microcontrollers with the multiply-accumulate, data massive, and deterministic software typically with zero overhead loops of the DSPs. With the inclusion of DSP functionality in microcontrollers many things such as modem support, compression of data voice, and image as well as filtering and speech synthesis/recognition are possible. This has been successfully done in some of the recently launched microcontrollers like Goal Semiconductor's VMX1016. This microcontroller possesses a powerful hardware arithmetic unit with 32-bit barrel shifter that can perform simultaneous 16-bit signed multiplication and 32-bit addition. Another example of this convergence is Hitachi's SH3-DSP based on a single instruction stream approach combining general purpose CPU and dedicated DSP functionality in a single architecture [21]. These types of microcontrollers facilitate upgradation of the embedded systems merely with the addition of software utilities.

Hidden Debugger

With the increasing design complexity debugging or verification has become the major bottleneck to meet the tight time to market schedule.

It is estimated that debugging time is almost 30–50% of the total development time. The conventional techniques such as extra design for testability by embedding test points or scan chains as well as external debugger are on the way of extinction. The new microcontrollers are struggling to sort out the resource requirement vs computational complexity dilemma. The use of logic analyzer for monitoring the clock quality as well as the bus activity has become very expensive and many times physically impossible with the shrinking footprint of microcontrollers operating at the clock speed as high as 100 Mhz. The IEEE-ISTO Nexus 5001 Forum which comprises around 25 major microcontroller manufacturers defines a common set of on-chip debug features, protocols, pins, and interfaces to external tools which may be used by real-time embedded application developers. As a result, more than 70% of leading embedded microcontroller vendors now have dedicated circuits and pins, whichassist in new product development based on the IEEE 1149.1 Joint Test Action Group (JTAG) 4-wire serial interface [22]. Apart from the JTAG interface, today's microcontrollers are enriched with the background debug mode (BDM) and on-chip emulation being achieved at higher abstraction level for reducing size and searching space in the debugging. Although the microcontroller community is still far from implementing the debugging solution for pointer problems, accessing unintialized memory, and interprocess interaction, the chips are dominated with single stepping and breakpoint utilities, e.g., the recently launched C8051F120 [23] has on-chip debug circuitry facilitating full-speed in-system debug features such as single-stepping, breakpoints, and modifying registers and memory.

Web-Enabled Microcontrollers

Embedded internetworking allows anytime, anywhere, control of the product with a regular self-maintenance and pervasive, wearable computing. The tangible benefits of Internet connectivity in products like air conditioner or washing machine are self-upgradation by downloading the software updates, lowering cost by fine-tuning energy consumption in peak hours, generating auto maintenance alarms, etc. Microcontroller manufacturers have come out with new processors having a built-in capability to cope with the challenges of web processing. Dallas Semiconductor has come out with the microcontrollers that can directly serve up web pages. TINI (Tiny InterNet Interface) [24] is a microcontroller-based development platform that executes code for embedded web servers. Remote devices can have preferences and settings adjusted from afar, just by having their administrator browse a web page hosted by the microcontroller no other computers required. The TINI development platform combines a powerful chipset and a Java runtime environment that exposes the extensive I/O capability of the Dallas microcontrollers. A Java programmer accesses the I/O from robust APIs (Application Program Interfaces) that include Ethernet, RS-232 serial, I2C, 1-Wire net, CAN, and memory-mapped parallel bus. By using these APIs, programmers code functions without worrying about the underlying interface to hardware peripherals. The runtime environment is tightly coded for optimized network communications and efficient device I/O throughput.

Dominance of Soft IP cores/Free Microcontroller?

The new VLSI devices such as FPGAs offer several advantages over microcontrollers. FPGA-based platform enables the developer to test (thanks to the JTAG philosophy) and add features in parallel without the need for repeating the complete testing of the platform once again. Increasing number of manufacturers are now offering the FPGA-based microcontroller core. The list of FPGA-based soft IP cores for 8051-based microcontrollers is available at the web page of Keil (http://www.keil.com/dd/ipcores.asp). For example, the C-8051 core is the HDL model of the Intel 8-bit 8051 microcontroller by Aldec Inc. [25]. The model is fully compatible with the Intel 8051 standard and is available in both EDIF and VHDL netlists formats. The EDIF netlist is used for the place and route process and VHDL is the post-synthesis netlist used for the simulation only. Both netlists are technology-dependent because they are created after the synthesis where the customer needs to specify a vendor, target family, etc. Even the source code of a set of HDL testbenches for the cores is also available.

1.8 Jump Starting Your First Embedded System Project

Figure 1.2 shows a minimum setup required to build microcontroller-based embedded system project. The foremost requirement is a low-end PC pre-loaded with the IDE to facilitate the code development, simulation and testing before actual dumping the code in the flash of the microcontroller. The Keil IDE is covered in depth in Chapter 2. Most of the projects developed in this book have been tested on AT89S52 which has the following features:

- 8K Bytes of In-System Reprogrammable Flash Memory
- Fully Static Operation: 0 Hz to 33 MHz

Figure 1.2 Minimum setup recommended for your Embedded System Laboratory

- Three-level Program Memory Lock
- 256 × 8-bit Internal RAM
- 32 Programmable I/O Lines
- Three 16-bit Timer/Counters
- Eight Interrupt Sources
- Programmable Serial Channel
- Low-power Idle and Power-down Modes
- 4.0V to 5.5V Operating Range
- Full Duplex UART Serial Channel
- Interrupt Recovery from Power-down Mode
- Watchdog Timer

- Dual Data Pointer
- Power-off Flag
- Fast Programming Time
- Flexible ISP Programming (Byte and Page Mode)

1.9 Execution of Embedded System Project: Is it a Gordian's Knot?

Here is a piece of thought for the student community as well as entrepreneurs looking for a successful embedded system project.

Project Idea to Execution

For successful project development, first you must develop an idea. There are lots of sources for idea generation. Ideally your ideas should move from research laboratories towards the marketplace. Many devices and processes around can be improved by the inculcation of microcontrollers. The automation of the nylon rubber stamp making machine described in the case studies is the best example. The thinking should ideally go on the following lines: How can the process be controlled using a microcontroller, ultimate efficiency, or throughput improvement with embedding of microcontrollers? Are there any simulation tools to model and estimate before the actual experimentation? In what way can the process be made more intelligent (or thinking) by using the microcontrollers? For answering the automation of a process plant using microcontroller, one must think about the on-chip resources and their effective usage (e.g., interrupt level/edge, usage of ports, timer/counter, and so on). The answer to the second question should throw light on the benefits of using the microcontroller in your project. (Can you think of green house controller without microcontroller? Or the amount of energy saved with the microcontroller-based corridor lighting which is implemented in the case studies.) According to the US Department of Energy [26], the electricity used to light businesses represents 25% of the energy they spend; so it is important to develop energy-saving devices. Even a simple switch equipped with photosensor and microcontroller can help to reduce lighting energy consumption by 30%.

Simulation is the best methodology to avoid project failure as well as to work in a time-efficient manner. It is always a good idea to simulate the things using the microcontroller IDE instead of rushing to the project board and actually dumping the code in the on-chip memory.

Moreover, there are simulation tools even to model physical systems (such as Proteus) which can be used together with the IDE to work out your project. The last question is little challenging. In order to make the microcontroller think or make it intelligent you have to resort to some novel items such as neural net implementation, extrapolation, and statistical techniques.

As the sole success of the embedded system project is on firmware, the software and the underlying hardware should not be treated as separate entities. Therefore, the term "Hardware–Software Co Design" has gained lots of significance in the field of embedded systems. Following issues pertaining to hardware and software are of utmost importance while executing any embedded system project.

Hardware and Software Issues

Even the simplest things dictate the final specifications of the final product. There is often a major gap between the theory of design and what "plays" in the real world. The most casual thing for a designer in any microcontroller-based product is the value of the crystal frequency which is 11.0592 MHz. The rationale behind this value is the ease of frequency division to yield exact clock rates for standard baud rate generation for the UART. However, in the applications where the serial communication is not at all in picture, the designer has lot of flexibility in choosing the crystal value. The crystal value decides the execution speed, e.g., Intel 8051 microcontrollers require a minimum of 12 oscillator cycles. This means if the crystal is of 12 MHz, then 8051 microcontroller works at performance of 1 MIPS. If a thorough analysis of the occurrence of real-time events reveals that this speed is not required, then a designer may go for a lower value of the crystal. The ultimate advantage is a significant reduction in power consumption. In CMOS-based versions of 8051, a linear relation between oscillator frequency and power consumption exists. Another lower range crystal value is 7.3728 MHz which can be used even for standard baud rate generation. The notable thing is: using counter1 of 8051; this gives an even 38.4 Kbps rate exactly, which is not possible even with 11.0952 MHz Xtal. The lower crystal value also enables to access low-speed peripherals and frees the system from electromagnetic interference (EMI) evident in high clock speed MCUs.

Problems such as reset, latch-up, memory corruption, and code runaway are found to fail the embedded system application due to ESD and EMI. Areas of MCUs typically vulnerable to ESD and EFT stresses include: power and ground pins, edge sensitive digital inputs, high frequency digital inputs, analog inputs, clock (oscillator) pins, substrate injection, general purpose I/O (GPIO) with multiplexed pin functions, etc. These kind of problems can be solved by using the microcontroller having proper package style, footprint, and maximum number of supply and ground pins [27].

While developing the software it is important to have a modular approach. The interrupt service routines should be as short as possible to reduce the interrupt latency. Readability and debugging specifications not only enhances the value of the software but also frees the novice to minimize the learning curve. Choice of the language plays an important role to decide the lead time, life and processor migration of your microcontroller-based product. Using the higher level languages like 'C' for writing programs offers some unique advantages such as faster development, improved portability, reusability, platform independence – all at reduced cost. Complex algorithms can be very easily implemented in 'C' rather than assembly. As the design evolves in due course, the required restructuring of the program without breaking the existing design can be easily done in 'C'. Although assembly language is the best choice for time critical programs, the user face lot of difficulties in managing large programs especially in memory allocation. "Porting" the assembly language code to another processor family is almost impossible or very difficult.

Chapter 2

Integrated Development Environment

Integrated development environment popularly known as IDE is a suite of software tools that facilitates microcontroller programming. The Keil IDE enables the embedded professional to develop the program in C and assembly as well. The IDE passes through the source code to check the syntax. The compilation leads to a hex file to be dumped in the microcontroller on-chip ROM. A quick session of simulation and debugging using the IDE ensures the working of the program beforehand. The user can verify the results as the package presents screenshots of on-chip resources. This chapter presents in-depth discussion on using the μVision 2 package of Keil IDE on MS Windows platform. It is recommended that while going through the discussion the user should access the μVision 2 package of the Keil. A step-by-step working as discussed in this chapter will empower the user to get familiar with the Keil IDE.

2.1 Getting Familiar with the IDE

The microcontroller product development cycle consists of several steps such as

- Development of code either in assembly or C
- Simulation of the code
- Dumping the code in microcontroller
- Prototyping or debugging if required by using in-circuit testing
- Emulation of the code in case of big project
- Refining the code, reprogramming and final testing

A microcontroller-based project generally makes several iterations through the above-mentioned steps before it sees the light of the day.

J.S. Parab et al. (eds.), Exploring C for Microcontrollers, 19–28.

Almost all the microcontroller manufacturers have come out with the development tools for their products. A suite of such software tools for microcontroller application development is refereed to as IDE. Some examples of the popular IDEs apart from Keil are MPLAB from Microchip, PE micro from Motorola, and AVR studio from Atmel.

Any typical IDE comprises many subcomponents such as a text editor for entering the code, building tools for converting the code into machine level, compiler to convert 'C' code to assembly or hex format, and linker to provide absolute addresses to the relocatable object code. The IDE is generally equipped with powerful code simulator that models operation of the target microcontroller as code execution is in process. IDEs are available as a software suite which runs on a stand-alone PC and displays and allows to modify the contents of virtual I/O and on-chip resources such as registers. Single-stepping and breakpoint facilities are provided to systematically execute and watch the cause/effect relationship of the code. A provision of instruction cycle calculation/display is also provided so as to see the time/memory efficiency of the code.

2.2 Working with Keil IDE

As the entire software development described in this book is based on the Keil 8051 Development Tools, it is worthwhile to study the tool in depth. The Keil IDE is a user-friendly software suite to develop your 'C' code and test the same in a virtual 8051 environment. The main feature of the Keil is that it allows C programmers to adapt to the 8051 environment quickly with little learning curve.

It offers the designer a device database of MCS-51 family from which the target device of interest can be chosen. The µVision IDE sets the compiler, assembler, linker, and memory options for the designer. The suite comes with numerous example programs to assist the designer to start his project. With the virtual environment, the available on-chip resources of the microcontroller chosen can be seen working on the PC screen. The simulation window facilitates very realistic simulation of both CPU and embedded peripherals. The graphical window shows the state and configuration of the embedded peripherals and displays the interaction of the microcontrollers with external peripherals. Although the simulation exercise consumes time, it helps to save the bugs and project failure in the long run.

A natural question that occurs to designer is in what way the conventional C programming for PC is different from the C programming with Keil or C51 programming. The basic objective of the conventional

C programming language was to make it work on the PC architecture with a high level interface. Therefore, it does not support direct access to registers. However, unlike the conventional C programming on PC, most of the microcontroller-based embedded systems applications demand reading and setting or resetting of single bits, or Boolean processing. The second main difference is the programming environment. Conventional C operated under the umbrella of an operating system may be Linux or Windows, wherein there is a provision of system calls the program may use to access the hardware. Almost all the programs written with microcontroller are for direct use where the while(1) loop is common. The above-mentioned differences are bridged in Keil compiler by adding several extensions to the C language, e.g., connecting of interrupt handlers. A good web-resources are available on this topic at www.massey.ac.nz/~chmessom/Chapter%206%20C%20ProgrammingFinal.pdf

2.3 Development Flow for the Keil IDE

The evaluation version of the Keil IDE for MCS-51 family (also known as C51 evaluation software) can be downloaded from the website www.keil.com. One can go up to a limit of 2 Kbytes of object code with the evaluation version. Once the Keil IDE is installed, a short cut appears on the desktop.

Step 1: Interfaces Offered by Keil IDE

After clicking on the above shown shortcut the following steps should be carried out. The first blank window will appear as shown in Figure 2.1.

You will be working with three windows presented by the Keil IDE. The first is target toward the extreme left of the screen which is blank at the moment, but will be updated as you go on working with the project. The source file, register header file, and the target microcontroller chosen is displayed here. The bookmarks for the user such as data sheet, user guides, library functions, etc., forms a ready reckoner for the developer. The program window which occupies most of the size of your screen displays the source code in C. At the lowermost end of the screen, an output window is presented which gives information regarding the errors and other output messages during the program compilation.

Open a new text file for writing your source code. This file has to be saved with .c extension. This source code file will be ultimately added

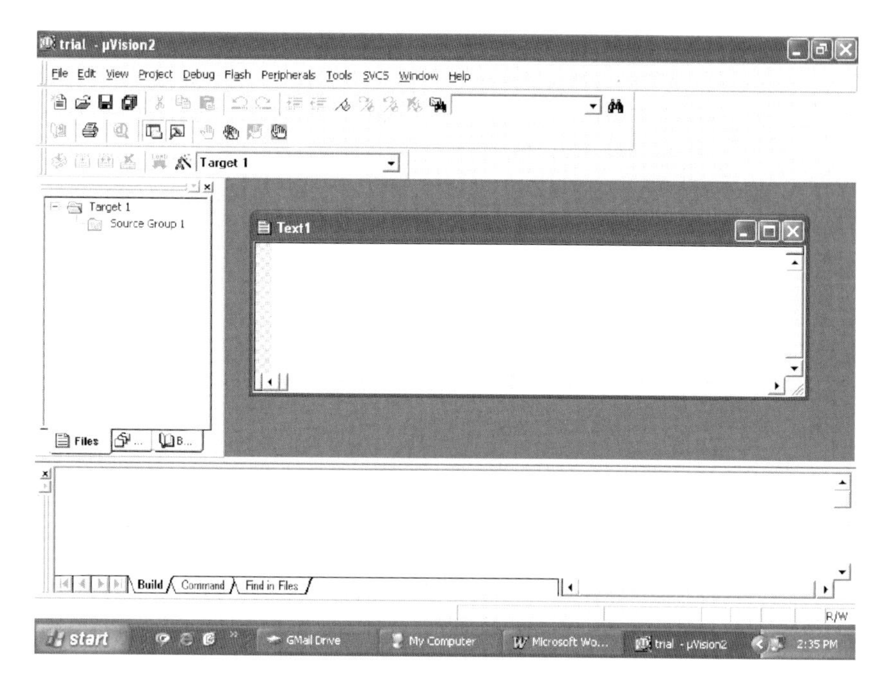

Figure 2.1 The starting interface of Keil IDE

into your project to be opened as per step 2. The addition should be done as per the instructions given in step 5.

Step 2: Opening a New Project

Here you can give a meaningful name for your project. After saving it will create a folder which will store your device information and source code, register contents, etc. Figure 2.2 shows a project opened with a name trial.

Step 3: Selecting a Device for the Target

After completing step 2, Keil will give an alert to select the device. The µVision 2 supports 45 manufacturers and their derivatives. In the exercise given in this book we have selected Atmel's 89S52 microcontroller as a target.

Step 4: Copying Startup Code to Your Project

The "startup.a51" will be added automatically to your project from the Keil library "c:\keil\c51\lib" to [c:\keil\c51\EXAMPLES\HELLO\ctrial\STARTUP.A51].

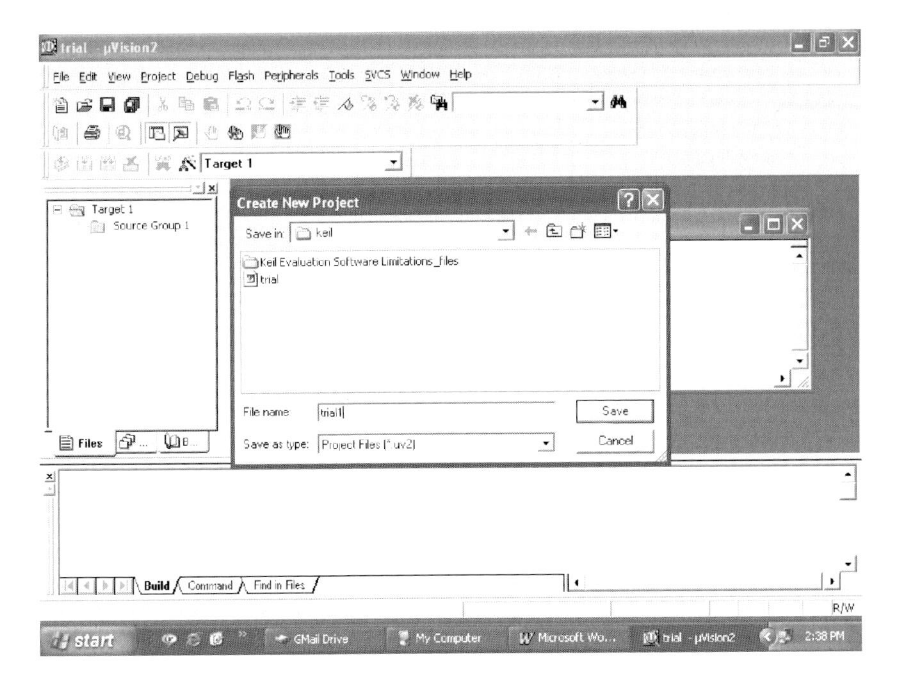

Figure 2.2 Opening a new project

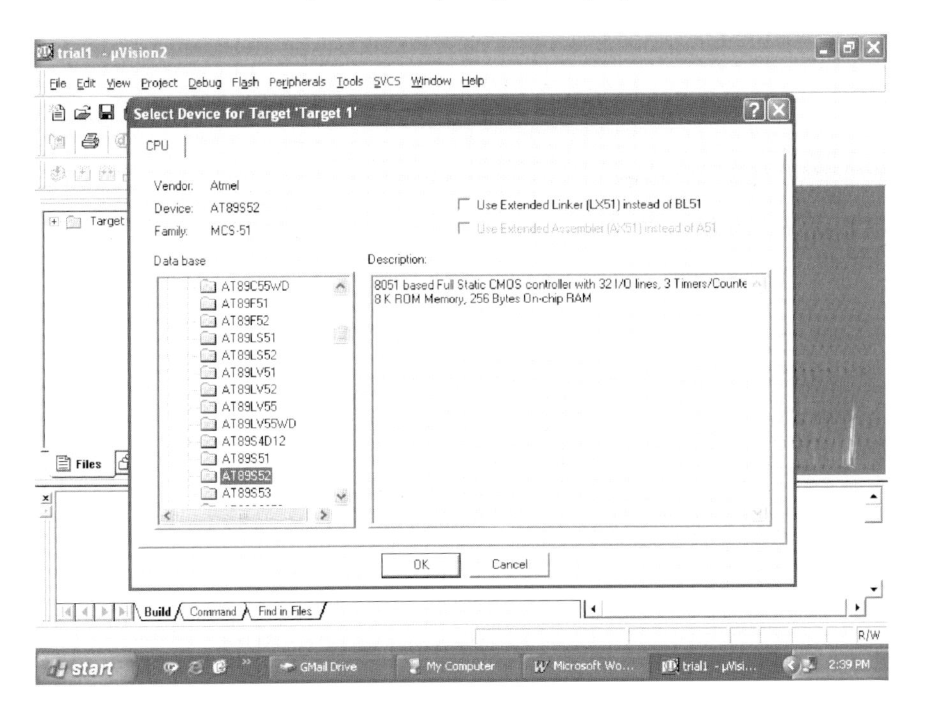

Figure 2.3 Selecting a device for the target

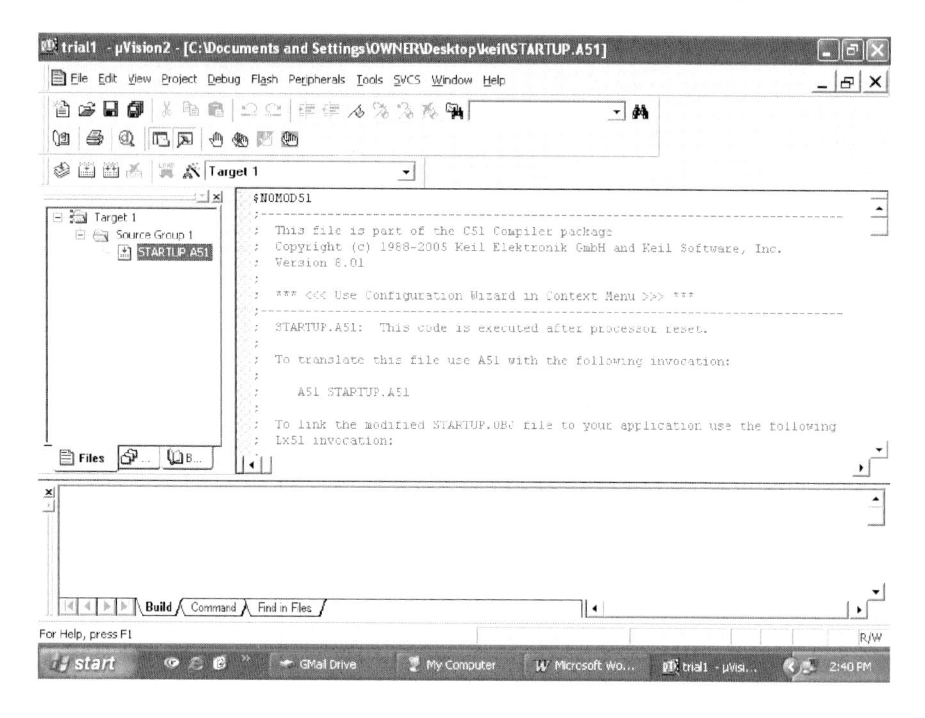

Figure 2.4 Copying startup code to your project

Step 5: Adding Your Program Source Code

To accomplish this, follow the steps:

- Right click "source group 1" followed by
- Choose "Add Files to Group 'Source Group 1'"
- Set "Files of type" to "All files (*.*)"
- Select "Startup.a51"

Observe the files getting updated in the target window. You will have to double-click on the C source file name displayed in the target window to view it.

Step 6: Configuring and Building the Target

Right click on target 1 in the target window, select the option for target 1, a window to choose the options for the target will be displayed. Here you can choose the microcontroller frequency, listing of files, output in hex, debug information, etc.

The important point here is choosing the appropriate memory model.

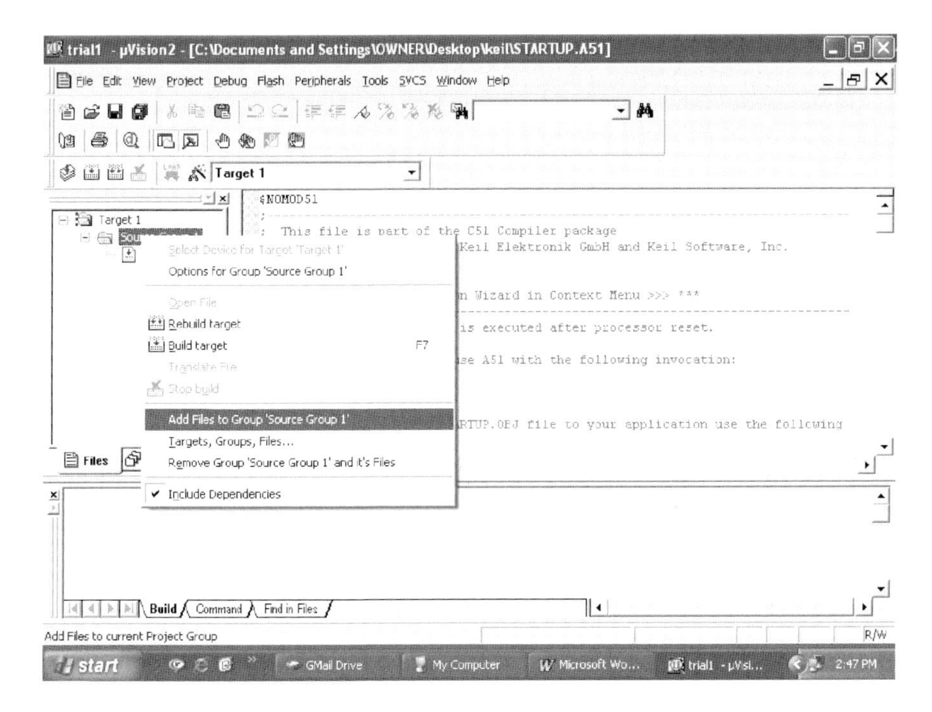

Figure 2.5 Adding your program source code

As per the on-line Keil IDE manual [28] C51 currently supports the following memory configurations:

- ROM: currently the largest single object file that can be produced is 64 K, although up to 1 MB can be supported with the BANKED model described below.
- All compiler output to be directed to EPROM/ROM, constants, look-up tables, etc., should be declared as "code".
- RAM: There are three memory models, SMALL, COMPACT, and LARGE
- SMALL: all variables and parameter-passing segments will be placed in the 8051's internal memory.
- COMPACT: variables are stored in paged memory addressed by ports 0 and 2. Indirect addressing opcodes are used. On-chip registers are still used for locals and parameters.
- LARGE: variables etc. are placed in external memory addressed by @DPTR. On-chip registers are still used for locals and parameters.

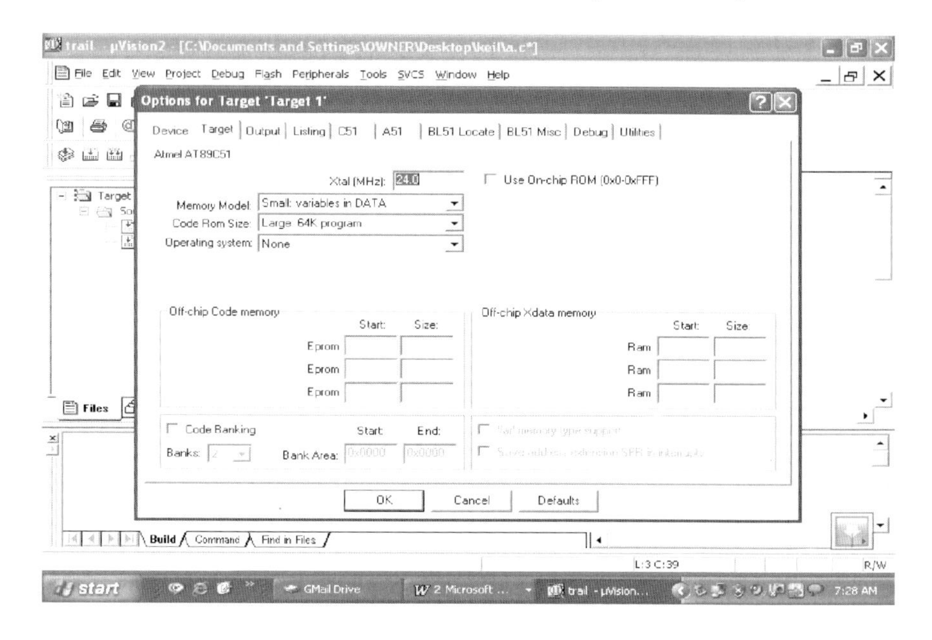

Figure 2.6 Configuring and building the target

Table 2.1 Choosing the best memory model for your C51 program

Model	*RAM supported*	*Best for*	*Worst for*
Small	Total RAM 128 bytes (8051/31)	Code size up to 4 K	Global variable (be kept minimum)
Compact	256 bytes off-chip, 128 or 256 bytes on-chip	High stack usage, programs with large number of medium speed 8-bit variables	Rarely used in isolation, usually combined with the SMALL switch reserved for interrupt routines.
Large	64 KB, 128, or 256 bytes on-chip	Easiest model to use	Never used in isolation, combined with small and compact

- BANKED: Code can occupy up to 1 MB by using either CPU port pins or memory-mapped latches to page memory above $0 \times$ FFFF. Within each 64 KB memory block a COMMON area must be set aside for C library code. Inter-bank function calls are possible.

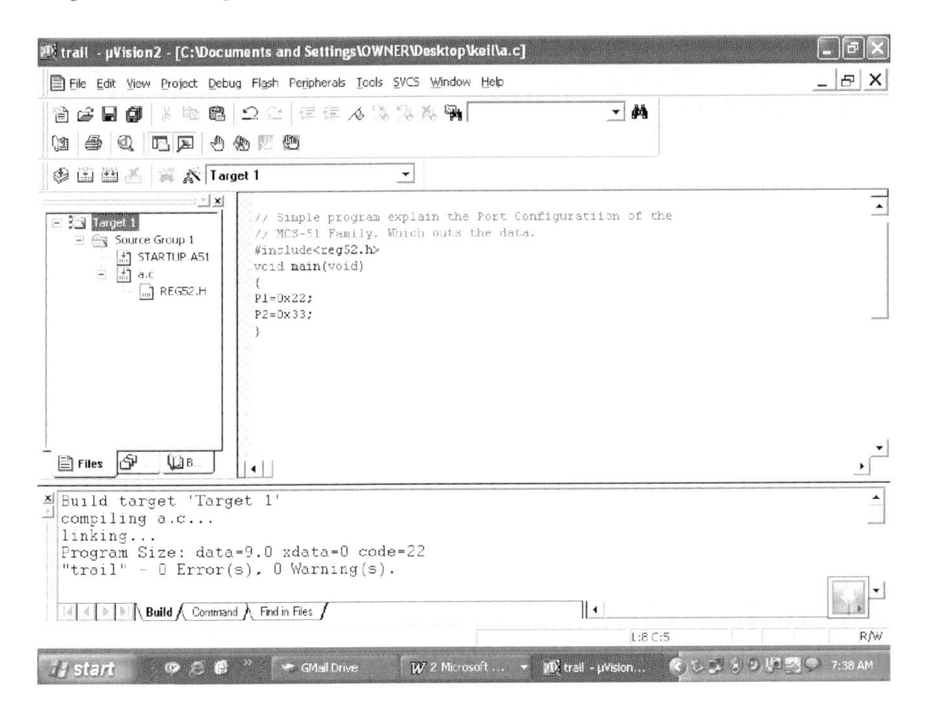

Figure 2.7 Compiling your program by pressing F7

Step 7: Compile Your Program by Pressing F7

Either press F7 or click on build target in the target window to compile your program. The output window will show the errors or warnings if any. You can also see the size of data, code, and external data, which is of immense importance since your on-chip memory is limited.

Step 8: Working in Simulated Mode

Once the program is successfully compiled, you can verify its functionality in the simulated mode by activating the debug window. For this press CTRL + F5 or go to the menu option "Debug" and select "Start and Stop Debug Section". Press F11 for single stepping or F5 for execution in one go. Go to the menu item "Peripheral" and select the appropriate peripherals to view the changes as the program starts executing.

Terminating the debug session is equally important. Click on "stop running" or ESC key to halt the program execution. You can make the changes to the program after coming out from the debug session

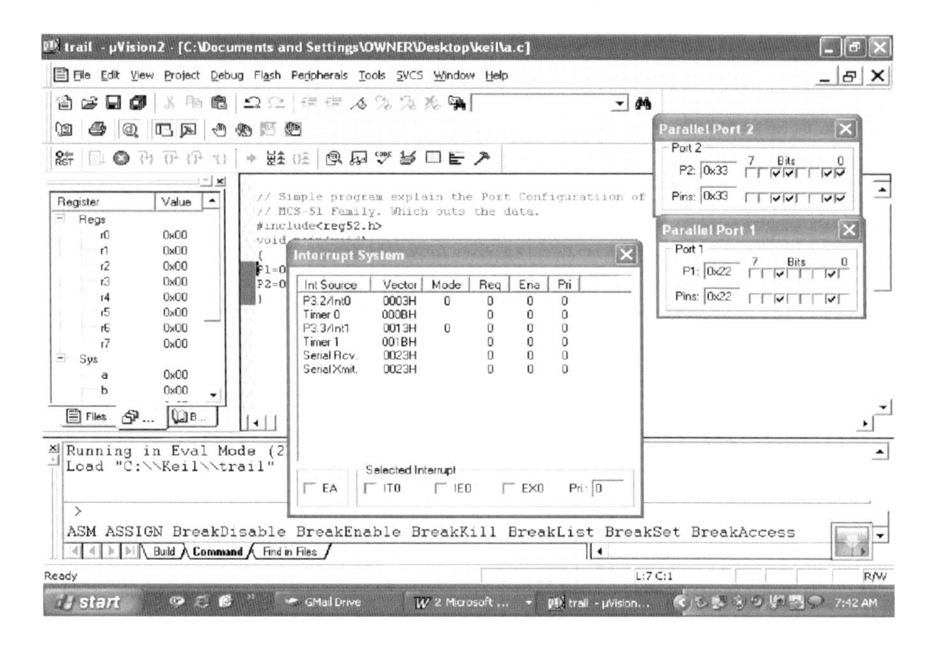

Figure 2.8 Simulation mode of Keil

by pressing start/stop debug session. Few iterations through the above-mentioned process will make your program completely bug-free and save your time and other resources on the actual hardware.

Step 9: Actual Dumping of the Code in Microcontroller's On-chip Memory

There are different mechanisms for this. Some programmers use ready-made programmers to program the microcontroller. If you are interested in purchasing one for your project, the list is enclosed. Design of various programmers has also been given at Wichit Sirichote's webpage at http://www.kmitl.ac.th/∼kswichit%20/.

Chapter 3

Art of C Programming for Microcontrollers

There are large number of textbooks and data sheets which give an overview of 8051 architecture, pinout, and other peripheral details. However, with the Keil IDE, you can take the benefit of the header files and need not remember the details like the port address, timer counter addresses, etc. The CA51 Compiler Kit for the 8051 microcontroller family supports all 8051 derivatives including classic devices and IP cores from companies like Analog Devices, Atmel, Cypress Semiconductor, Dallas Semiconductor, Goal, Hynix, Infineon, Intel, OKI, Philips, Ailicon Labs, SMSC, STMicroelectronics, Synopsis, TDK, Temic, Texas Instruments, and Winbond.

3.1 Familiarizing with Your Compiler Capabilities

The C for microcontrollers and the standard C syntax and semantics are slightly different. The former is aimed at the general purpose programming paradigm whereas the latter is for a specific target microcontroller such as 8051. The underlying fact is that everything will be ultimately mapped into the microcontroller machine code. If a certain feature such as indirect access to I/O registers is inhibited in the target microcontroller, the compiler will also restrict the same at higher level. Similarly some C operators which are meant for general purpose computing are also not available with the C for microcontrollers. Even the operators and constructs which may lead to memory inefficiency are not available in C programming meant for microcontrollers. Be aware that the target code should fit in the limited on-chip memory of the processor. Even the I/O functions available in standard C such as printf() or scanf() are either not made available in C compilers for microcontrollers or advised not to use them. These functions eat up lot of memory space and are not time-efficient owing to the dragging of supporting functions like floating point routines and lot of delimiters. Another striking difference in case of embedded systems programs is that they do not

J.S. Parab et al. (eds.), Exploring C for Microcontrollers, 29–36.

have the umbrella or support of the operating system. The programmer has to be accustomed with the absence of system calls which makes life easy in traditional C.

The C for microcontrollers is more closer to the assembly language than the conventional or traditional C. Few examples to support the above statement are presence of arithmetic/logic/bitwise operations to resemble machine operations, Use of pointers which corresponds to memory access at machine level etc. The detailed information regarding the features of 'C' supported may be found in the document C×51 Compiler User's Guide for Keil Software [28].

3.2 Whether to Use Headers or Not?

As a professionally embedded systems developer one can write a standalone 'C' code, requiring no header files, and it will compile correctly with the Keil C51 compiler. However, this requires a through study of the target microcontroller architecture. With the large number of derivatives of the 8051, it is difficult to remember the on-chip resources, e.g., the MCS-51 versions from Atmel such as AT89C2051 or AT89C51 have two timers while the other members such as At89C52 or AT89S53 have three timers. Therefore, although there is little memory space overhead, it is advisable to use the header files which defines all the on-chip resources for the target microcontroller, e.g., whenever you choose the target device such as AT89C52, Keil IDE gives the information regarding all its on-chip resources in the window at the right side as shown in Figure 3.1.

You can observe another advantage of working with the header file. Define the header file <REG51.h> or <REG52.h> in the program window and select the target device as 80C51. Now open the header file, which will reflect all the on-chip resources of 80C51 as shown below:

```
/*--------------------------------------------------
REG51.H

Header file for generic 80C51 and 80C31 microcontroller.
Copyright (c) 1988-2002 Keil Elektronik GmbH and Keil Software, Inc.
All rights reserved.
--------------------------------------------------*/

#ifndef __REG51_H__
#define __REG51_H__

/* BYTE Register */
sfr P0 = 0x80;
sfr P1 = 0x90;

/* BIT Register */
/* PSW */
sbit CY = 0xD7;
sbit AC = 0xD6;

sbit EX1 = 0xAA;
sbit ET0 = 0xA9;
sbit EX0 = 0xA8;

/* IP */
sbit PS = 0xBC;
```

```
sfr P2 = 0xA0;
sfr P3 = 0xB0;
sfr PSW = 0xD0;
sfr ACC = 0xE0;
sfr B = 0xF0;
sfr SP = 0x81;
sfr DPL = 0x82;
sfr DPH = 0x83;
sfr PCON = 0x87;
sfr TCON = 0x88;
sfr TMOD = 0x89;
sfr TL0 = 0x8A;
sfr TL1 = 0x8B;
sfr TH0 = 0x8C;
sfr TH1 = 0x8D;
sfr IE = 0xA8;
sfr IP = 0xB8;
sfr SCON = 0x98;
sfr SBUF = 0x99;

sbit TB8 = 0x9B;
sbit RB8 = 0x9A;
sbit F0 = 0xD5;
sbit RS1 = 0xD4;
sbit RS0 = 0xD3;
sbit OV = 0xD2;
sbit P = 0xD0;

/* TCON */
sbit TF1 = 0x8F;
sbit TR1 = 0x8E;
sbit TF0 = 0x8D;
sbit TR0 = 0x8C;
sbit IE1 = 0x8B;
sbit IT1 = 0x8A;
sbit IE0 = 0x89;
sbit IT0 = 0x88;

/* IE */
sbit EA = 0xAF;
sbit ES = 0xAC;
sbit ET1 = 0xAB;
sbit TI = 0x99;
sbit RI = 0x98;
sbit PT1 = 0xBB;
sbit PX1 = 0xBA;
sbit PT0 = 0xB9;
sbit PX0 = 0xB8;

/* P3 */
sbit RD = 0xB7;
sbit WR = 0xB6;
sbit T1 = 0xB5;
sbit T0 = 0xB4;
sbit INT1 = 0xB3;
sbit INT0 = 0xB2;
sbit TXD = 0xB1;
sbit RXD = 0xB0;

/* SCON */
sbit SM0 = 0x9F;
sbit SM1 = 0x9E;
sbit SM2 = 0x9D;
sbit REN = 0x9C;

#endif
```

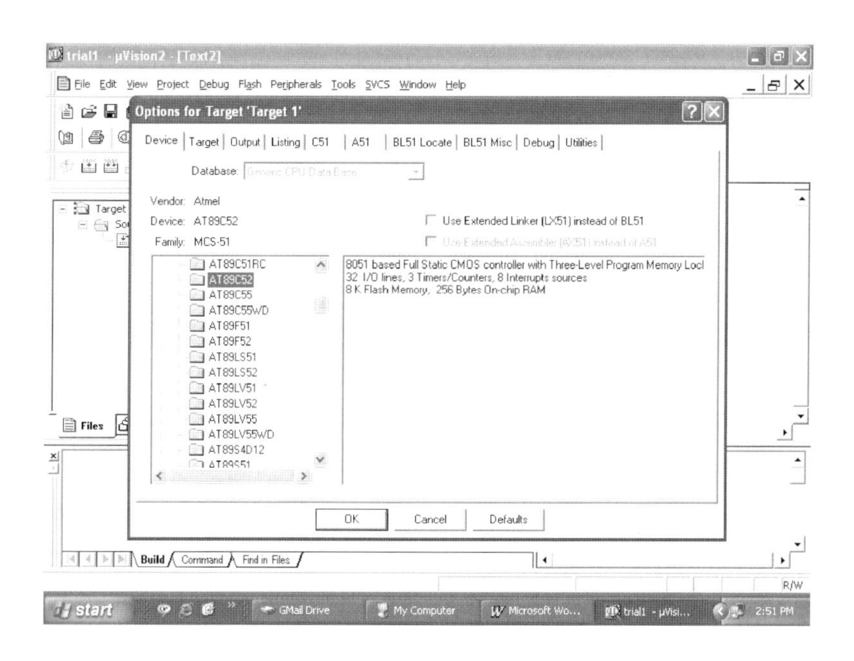

Figure 3.1 On-chip resources in the window

Now select another target device 80C52. This device has three timers. Now the header file will automatically update the database of on-chip resources pertaining to this device as shown below.

```
/*-----------------------------------------------------

REG52.H

Header file for generic 80C52 and 80C32 microcontroller.
Copyright © 1988–2002 Keil Elektronik GmbH and Keil Software, Inc.
All rights reserved.

------------------------------------------------------*/

#ifndef __REG52_H__
#define __REG52_H__

/* BYTE Registers */
sfr P0 = 0x80;
sfr P1 = 0x90;
sfr P2 = 0xA0;
sfr P3 = 0xB0;
sfr PSW = 0xD0;
sfr ACC = 0xE0;
sfr B = 0xF0;
sfr SP = 0x81;
sfr DPL = 0x82;
sfr DPH = 0x83;
sfr PCON = 0x87;
sfr TCON = 0x88;
sfr TMOD = 0x89;
sfr TL0 = 0x8A;
sfr TL1 = 0x8B;
sfr TH0 = 0x8C;
sfr TH1 = 0x8D;
sfr IE = 0xA8;
sbit EX0 = IE^0;

/* IP */
sbit PT2 = IP^5;

sfr IP = 0xB8;
sfr SCON = 0x98;
sfr SBUF = 0x99;

/* 8052 Extensions */
sfr T2CON = 0xC8;
sfr RCAP2L = 0xCA;
sfr RCAP2H = 0xCB;
sfr TL2 = 0xCC;
sfr TH2 = 0xCD;

/* BIT Registers */
/* PSW */
sbit CY = PSW^7;
sbit AC = PSW^6;
sbit F0 = PSW^5;
sbit RS1 = PSW^4;
sbit RS0 = PSW^3;
sbit OV = PSW^2;
sbit P = PSW^0;
//8052 only
sbit RCLK =
T2CON^5;
sbit TCLK =
T2CON^4;

/* TCON */
sbit TF1 = TCON^7;
sbit TR1 = TCON^6;
sbit TF0 = TCON^5;
sbit TR0 = TCON^4;
sbit IE1 = TCON^3;
sbit IT1 = TCON^2;
sbit IE0 = TCON^1;
sbit IT0 = TCON^0;

/* IE */
sbit EA = IE^7;
sbit ET2 = IE^5;
//8052 only
sbit ES = IE^4;
sbit ET1 = IE^3;
sbit EX1 = IE^2;
sbit ET0 = IE^1;
```

```
sbit PS = IP^4;
sbit PT1 = IP^3;
sbit PX1 = IP^2;
sbit PT0 = IP^1;
sbit PX0 = IP^0;

/* P3 */
sbit RD = P3^7;
sbit WR = P3^6;
sbit T1 = P3^5;
sbit T0 = P3^4;
sbit INT1 = P3^3;
sbit INT0 = P3^2;
sbit TXD = P3^1;
sbit RXD = P3^0;

/* SCON */
sbit SM0 = SCON^7;
sbit SM1 = SCON^6;
sbit SM2 = SCON^5;
sbit REN = SCON^4;
sbit TB8 = SCON^3;
sbit RB8 = SCON^2;
sbit TI = SCON^1;
sbit RI = SCON^0;

/* P1 */
sbit T2EX = P1^1;
//8052 only
sbit T2 = P1^0;
//8052 only

/* T2CON */
sbit TF2 = T2CON^7;
sbit EXF2 =
T2CON^6;
sbit EXEN2 =
T2CON^3;
sbit TR2 =
T2CON^2;
sbit C_T2 =
T2CON^1;
sbit CP_RL2 =
T2CON^0;

#endif
```

Note the bold letters which are exclusively added for the 80C52 only.

In nutshell the first program sentence will be always a statement **#include <reg51.h>** for MCS-51 family. The standard initialization and startup procedures for the 8051 are contained in startup.a51. This file is included in your project and will be assembled together with the compiled output of your C program. For custom applications, this startup file might need modification, because the stack or stack space, etc., are predefined as per the memory model.

3.3 Basic C Program Structure

Listing below shows a basic structure for a C program.

```
//-----------------------------------
//Basic blank C program that does nothing
// Includes
//-----------------------------------
#include <reg51.h> // SFR declarations

void main (void)
        {

while(1);
              {
                    body of the loop              // Infinite loop
              }
        }                                  // match the braces
```

Note: Similar to conventional C, all the variables must be declared at the start of a code block, you cannot declare variables in the midst of the program statements.

3.4 Differences from ANSI C

There are marginal differences of Cx51 compiler from the ANSI C compiler. These differences are twofold, viz. compiler related and library related. The details are given in the help menu of the Keil IDE. Frequently used things are given below for ready reference of the programmers.

Variable Types

The Keil C compiler supports most ANSI C variable types and in addition has several of its own.

Standard Types

Only floating point variable types of ANSI C are not supported by Keil C compiler as the basic 8051 core is not efficient for the same. The types supported are as follows:

Type	*Bits*	*Bytes*	*Range*
Char	8	1	-128 to $+127$
Unsigned char	8	1	0 to 255
Enum	16	2	$-32,768$ to $+32,767$
Short	16	2	$-32,768$ to $+32,767$
Unsigned short	16	2	0 to $65,535$
Int	16	2	$-32,768$ to $+32,767$
Unsigned int	16	2	0 to $65,535$
Long	32	4	$-2,147,483,648$ to $2,147,483,647$
Unsigned long	32	4	0 to $4,294,697,295$

Keil Types

The bit and Boolean capabilities of MCS-51 family are well explored by the Keil C Compiler by adding several new types. Some of them are as follows:

Type	*Bits*	*Bytes*	*Range*
Bit	1	0	0 to 1
Sbit	1	0	0 to 1
Sfr	8	1	0 to 255
Sf16	16	2	0 to $65,535$

Of these, only the bit type works as a standard variable. The other three have special behavior that a programmer must be aware of. The bit type gets allocated to the 8051's bit-addressable on-chip RAM. Note that you cannot declare a bit pointer or an array of bits. The special function registers are declared as a sfr data type in Keil C. If you give the address of the SFR as 0x80h then it gets allocated to port 0 of 8051. Extensions of the 8051 often have the low byte of a 16 bit register preceding the high byte. Therefore the timer 0 for instance should be declared as a 16-bit special function register, sfr16, giving the address of the low byte:

```
sfr16 TMR0RL = 0x8A;          // Timer0 reload value
sfr16 TMR0 = 0x8B;       // Timer0 lower
```

Keil Variable Extensions

In consistent with the 8051 architecture Keil offers various types of variable extensions. Following table summarizes the same. Assembly language mnemonics are used to indicate the subsequent meaning of the variable extension used.

Extension	**Memory Type**	**Related ASM**
Data	Directly-addressable data memory (data memory addresses 0x00-0x7F)	MOV A, 056h
Idata	Indirectly-addressable data memory	MOV R0, #09Ah
	(data memory addresses 0x00-0xFF)	MOV A, R0
xdata	External data memory	MOVX @DPTR
Code	Program memory	MOVC @A+DPTR

In case the memory type extension is not specified, the compiler will decide which memory type to use automatically, based on the memory model (SMALL, COMPACT, or LARGE, as specified in the project properties in Keil).

Chapter 4

Exploring the Capabilities of On-Chip Resources Programming for I/O Ports, Interrupts and Timer/Counter

4.1 Importance of Ports

Ports are the means through which the microcontroller communicates with the outside world. They are the most common I/O devices. An interesting point to note is that the number of I/O pins or ports are decided by the packaging style of the microcontroller. A comparison of various packages is given below: A new trend in 8-bit controller design is to place the chip inside a 20-, 16-, or even 8-pin PDIP or SOP package, allowing for very small and cheap embedded systems to be constructed [29].

Fortunately most of the derivatives of the 8051 are generally offered in 40-pin DIP or 52-pin PLCC packages and there is not much variation in terms of package pinouts. The obvious advantage of the upward compatibility maintained into the different versions of the 8051, is the direct replacement of the latest advanced 8051 from Intel to its cheap predecessor without even changing the printed circuit board. This "drop-in" replaceability is rarely true of other microcontroller architectures [30].

Playing with the Ports

Program 4.1: Simulated Port Testing Using Keil

The program illustrates simple port programming in a simulated mode using the Keil compiler. A sample data is sent to the port and the output is observed in the simulator window. You may as well connect LEDs to observe the output.

J.S. Parab et al. (eds.), Exploring C for Microcontrollers, 37–67.

Table 4.1 Comparison of the packages

Package	*Number of pins*	*Features*	*Applications*
PDIP (plastic dual in-line package)	8–64	Through hole type devices, dense style package, pins/area ratio of PDIP not very efficient	Prototype or hobbyist type applications, smaller 8-bit microcontrollers
PLCC (plastic leaded chip-carrier)	20 up to 80+	Through hole type devices	Small, low-pincount microcontrollers
SOIC (small outline integrated circuit) or "SOP" (small outline package)	8–64 pins	Pins are spaced much closer together, package designed to be soldered directly to the top of a board	
QFP (quad flat package)	8–220+ pins	Sophisticated soldering workstation required to deal with the pins closely spaced	16- to 32-bit microcontrollers to 64-bit superscalar microprocessors

Program Source Code

```
**************************************************

#include<reg51.h>     /* special function register declarations for the
                         intended 8051 derivative. The header file
                         contains the definitions of special function
                         registers (SFRs) and other functions. */
void main(void)       // Start of the Main function.
{
        P1=0x44;      /*Send 44H to port P1*/
        P2=0x55;      /*Send 55H to port P2*/

}                     // End of the main function.

**************************************************
```

The screenshot of simulation window shown in Figure 4.1 the data delivery to port 1 and 2.

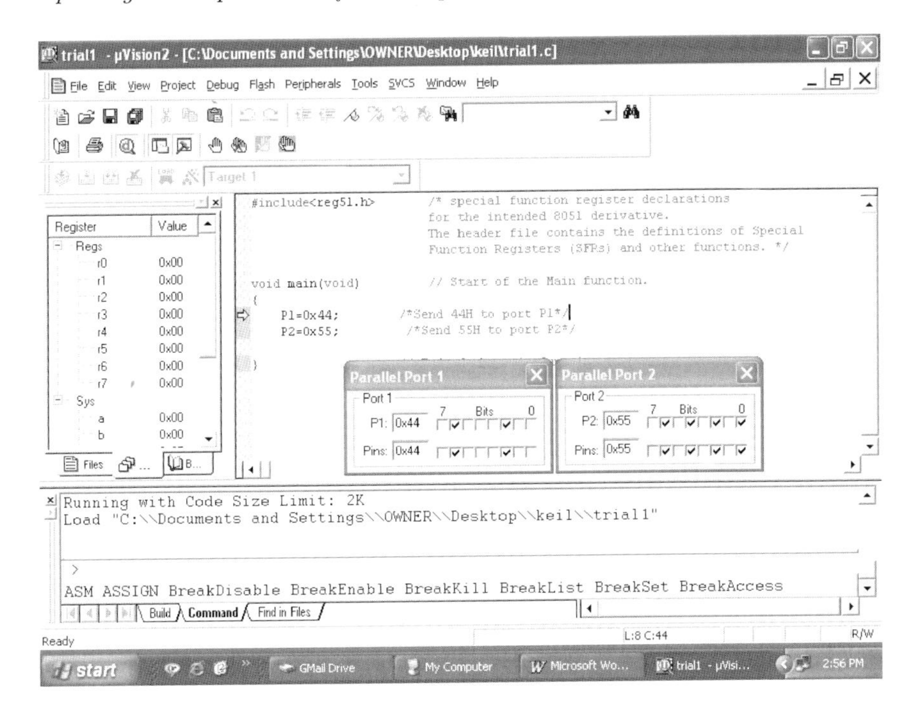

Figure 4.1 Screenshot of simulation window showing the ports

Program 4.2: Sending Out Different Contents to Port 1

Different data bytes are sent to port 1 after delay. The results can be checked in the simulation window, which shows the changing of data on the port 1 lines.

Program Source Code

```
*********************************************

#include <REG52.H>

void delay(void);

void delay(void)                    /* invoking delay function */
{
int i;
for (i=0;i<30000;i++);
for (i=0;i<30000;i++);
}
```

```
void main (void)                    // Starting of the main function.
{
        while(1)
        {
        P1 = 0x01;               // Out 01 on the port 1.
        delay();                 // Delay function.
        P1= 0x02;                // Out 02 on port1.
        delay();
        P1=0x04;
        delay();
        P1=0x08;
        delay();
        P1=0x10;
        delay();
        P1=0x20;
        delay();
        P1=0x40;
        delay();
        P1=0x80;
        delay();
        P1=0x40;
        delay();
        P1=0x20;
        delay();

        P1=0x10;
        delay();
        P1=0x08;
        delay();
        P1=0x04;
        delay();
        P1=0x02;
        delay();
        P1=0x01;
        delay();
        }                           // termination of the while loop.
}                                   // termination of the main function.
```

****

Program 4.3: Generating Square Wave on the Port

A number is assigned to port 1 which simulates input operation. The port 1 contents are complemented and sent to the port 1. The square wave can be observed on port 1 by using CRO.

Program Source Code

```
*********************************************


#include<reg51.h>

void delay(int);

void delay(int k)
{
int i,j;
for (j=0;j<k+1;j++){
for (i=0;i<100;i++);
} }

void main(void)                 // Start of the main function.

{
        int N1=0;               // Defining the integer value for the
number.
        while(1)                // Infinite loop
        {
          P1=N1 ;            /* Assign N1 to port P1*/
          N1 =~N1;              /*Complement N1*/
        delay(10);            /* Delay which defines the on and
                                off time for the Square wave. */
          P1 = N1;              /*Send it to port P1*/
          delay (10);

        }                       // End of while loop.

}                               // End of main.

*********************************************
```

Program 4.4: Sending Data to the Ports by Assigning Values to a Character

This program illustrates assigning values to the character and then sending it to the port. Here the characters a, b, c are assigned with the Hex numbers and these values are sent to ports.

Program Source Code

```
***********************************************
#include<REG51.H>

void main(void)
{
        char a,b,c;                 // Define the characters a, b and c.

        while(1)
        {
                a=0x01;             /* Assign 01H to 'a' */
                b=0x02;             /* Assign 02H to 'b' */
                c=0x03;             /* Assign 03H to 'c' */
                P0=a;               /* Send value assigned to 'a' to port 0*/
                P1=b;               /* Send value assigned to 'b' to port 1*/
                P2=c;               /* Send value assigned to 'c' to port 2*/
        }
}

***********************************************
```

Program 4.5: Configuring the Ports as Input

By default all the ports act as output port. They can be configures as input by writing FFH to them. The programmer need not bother about the actual SFR address. Thanks to the header file included at the top.

The actual input operation is simulated by assigning values to the variables. The actual DIP switch interfaced as input device is covered further in the text.

Program below illustrates the simplest way to define the ports as an input port and how to read the input from the ports and the result is observed in the simulation window.

Program Source Code

```
**********************************************************
#include<REG52.H>

void main(void)
{
	char N1,N2,N3;                  // Defining the characters.

	P0=0xFF;                        // set P0 as input
	P1=0xFF;                        // set P1 as input

	while(1)
	{
		N1=P0;                      /*Input value of P0 in N1*/
		N2=P1                       ;/*Input value of P1 in N2*/
		N3=N1+N2;                   // Addition of the two inputs
		P2=N3;                      /*Send value of N3 to port 2*/
	}
}

**********************************************************
```

Program 4.6: Software Delay Loops Combined with I/O Ports

The program demonstrates generation of software delay using two loops operating on a single count. An integer x is declared and assigned a count equal to 10,000. "While(1)" loop keeps the program in continues loop. The other two loops are conditional with the first loop executes when "while($x > 1$)" is valid, i.e., when x is greater than 1. The other loop executes when x is less than 10,000. Note the usage of decrement (−−) and increment (++) operators. Both these loops are embedded in the main loop.

To observe the output of this connect the port 1 to LEDs. The output can also be verified in the simulator window.

Program Source Code

```
**********************************************************
#include <reg51.h>

void main (void)
```

```
{
int x = 10000;

while(1)
{
    while(x > 1)              /* 10000 count delay */
      {
      P1 = 0x0f;              /* set leds on PORT1 to 00001111 */
      x--;                    /* decrement x               */
      } // end of first loop
    while(x < 10000)          /* 10000 count delay    */
      {
      P1 = 0xf0;              /* set leds on PORT1 to 11110000 */
      x++;                    /* increment x               */
      } // end of second loop.
  } // End of open loop.
} // End of main.
```

****

Program 4.7: Alternate Method of Software Delay Generation with Port Programming

This program illustrates the use of "FOR" construct for delay generation which toggles only one port pin (P1.1).

Note the use of ! operator for complementing the variable.

Program Source Code

****

```
#include<reg52.h>

sbit vin = P1^0 ;                 //Pin declaration vin = port 1 pin 0
void main(void)
  {
        int i;
        vin = 1;                            //Make port pin P1.0 high
        while(1)
        {
        for(i=0;i<=3000;i++);    // Delay generation
```

```
        for(i=0;i<=3000;i++);
        vin = !vin;                   //complement port pin P1.0 after
                                        delay
        }
  }
*******************************************************
```

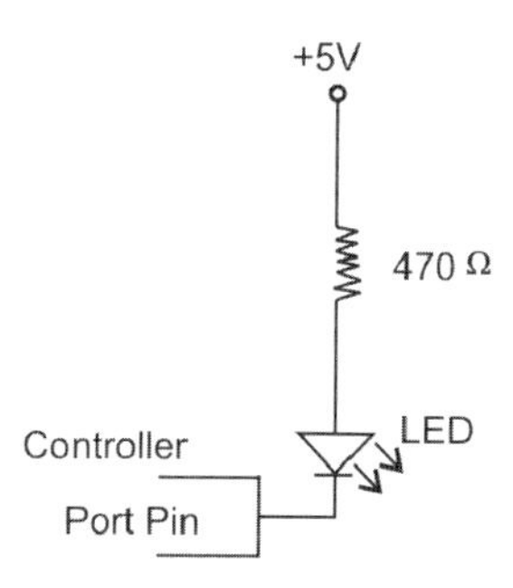

Figure 4.2 LED interface to microcontroller port pin

Program 4.8: Shifting the Port Contents to Left

This program shifts the contents of port 0 to left. The result may be seen in the simulation window by opening the menu option for peripherals. The port bits goes on changing one by one.

Program Source Code

```
*******************************************************
#include <REG52.H>

void main(void)
{
        int i;
        char N1;

        while(1)
        {
                N1=0x01;
                while(N1!=0x00)     //Continue the loop till N1 not
                                      equal to 00H
                                    //When N1=00H program exits
                                      the loop
                {
```

```
                for(i=1;i<10000;i++);       //delay
                P0=N1;                      //Send value of N1 to
                                              port 0
                N1=N1<<1;                   //Left shift N1 by one
                                              bit position
            }
        }
}           /* try right shift by using right shift operator '>>' */
```

****

4.2 Simple Ideas for Port Expansion

The users are worried when the number of I/O port lines poses a bottleneck for applications especially in the areas of robotics and industrial process control. In these types of application domains the user may either go for another microcontroller with more I/O pins or interface chips like PPI8255 to the existing ports. However, there are many simple methods to expand the port. Few of them tried by embedded system developers are given herewith the source URL, so that the user may adopt them for their applications.

- 2 Wire Input/Output for 8051 type CPUs [31].

  The application note explains the use I2C bus interfaced to shift registers for expansion of ports. 4094 shift registers have been used with the additional feature of tri-state output at power up until the CPU has written valid data, preventing power up "glitches".

- 8051 I/O expansion using common shift register chips [32].

  The application note gives complete schematic and assembly code to expand 8051 I/O using shift registers.

4.3 LED Interfacing

The simplest applications such as port testing or indication may only require on/off-type LEDs for displaying status information. For such applications, LEDs can be controlled directly by a general-purpose I/O pin capable of sourcing or sinking the necessary current required for lighting the LED, typically between 2 mA and 10 mA. LEDs have exponential voltage-current characteristics, that means around the operating point there is rapid increase in current for little increase in voltage. Therefore ideally a constant current source is desirable to drive the LEDs. However, the same can be simulated by using a series resistor connected to the port pin so that the voltage across the LED will be fairly constant.

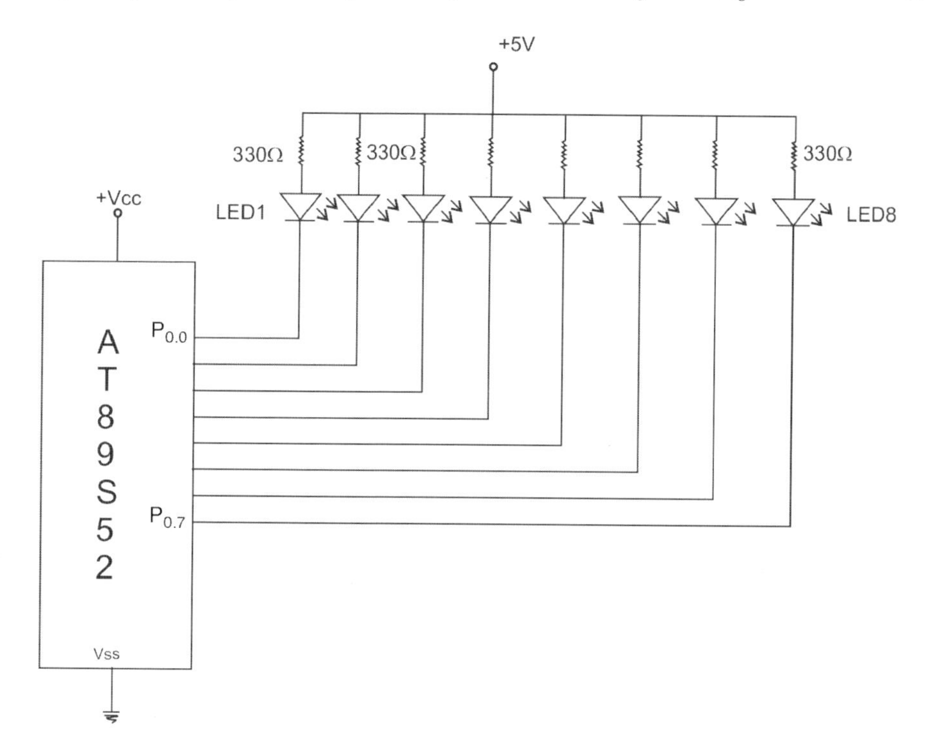

Figure 4.3 Interfacing diagram for a LED

Generally a common anode connection is chosen in a digital environment because pull-down is much easy as compared to push-up. If an external power supply is used, then the corresponding brightness will be much better as compared to driving logic 1 through the microcontroller port pins. From a programmer's point of view this means by writing logic '0' the LED will glow (see Figure 4.3).

Three things are required for working with the LEDs: the driving voltage, the current to be passed, and the forward voltage drop across the LED. For a blue LED with 12 V supply, in order to glow a single LED, a current of the order of 20 mA is required to be passed. With the forward voltage drop of 3.4 V, the value of the resistor can be calculated using simple Ohm's law as follows:

$$R = V_S - V_L / I$$

Where V_S = the supply voltage
V_L = voltage across the LED
I = Current through the LED.

The details of LED interfacing to microcontroller have been covered at several references. The microcontrollers I/O port pin is configured as an output. The choice of the series resistor and its wattage is covered in many standard books.

4.4 Relevance of LEDs in Today's Lightening Industry

Street lighting is fast emerging as a potentially strong market for LEDs, since these devices have reached a level of output and efficiency to make them viable replacements for incumbent lighting technologies. The adavntages of LED-based street lights are reduced energy consumption, less maintenance, and better light quality [33]. Novel interfacing techniques based on microcontroller are emerging which ranges from simple low-cost current limiting resistors to dynamic constant current utilizing PWMs and analog-to-digital converters. With HB-LED lighting requiring the use of low-voltage electronic control, low-cost microcontrollers are useful in bringing additional compelling features to household lighting. Microcontrollers are making it easier for lighting designers to incorporate a reliable connectivity interface such as DALI and DMX512 that will provide the backbone for adding remote user controls [34]. The above mentioned interfaces are based on packet-based system to deliver the data to the LED matrix. The technique has been covered in depth in this chapter.

4.5 Different Programs for LED Interfacing

Program 4.9: Blinking LED

This program illustrates simple LED blinking at a constant interval. A simple technique of complementing a bit is used with changing software delay. The hardware shown Section 4.3 is used.

Program Source Code

```
*****************************************************
#include <REG52.H>        /* special function register declarations
                          for the intended 8051 derivative */
void delay(void);         /*declaring a delay routine for blinking
interval*/

void delay(void)
{
int i;
```

```
for (i=0;i<30000;i++);        /* count */
for (i=0;i<30000;i++);
}

void main(void)               // invoking main function.
{
bit mybit;                    // define the bit as a mybit.

mybit = 1;                    // assign the mybit a logic high

while(1)                      // indefinite loop
{
mybit = !mybit;               // complement the bit
P1 = mybit;                   // send the complemented bit to port 1
delay();                      // invoke the delay routine
}                             // termination of the while loop
}                             // termination of the main function

***************************************************
```

Program 4.10: Scrolling LED

This program illustrates the interfacing of LED with scrolling pattern. The display appears as if the LED is shifting. The LED connected to the port 0 glows one by one. A software delay is used to define scrolling interval.

Program Source Code

```
***************************************************
#include <REG52.H>

void main(void)
{
        int i;                        // define the integer number
        char N1;                      //define the character to assign the
                                      hex number
        while(1)
        {
                N1=0x01;
                while(N1!=0x00)       // routine for scrolling interval
                {
                        for(i=1;i<10000;i++);
```

```
                P0=N1;        // display the LED corresponding
                              to N1
                N1=N1>>1;     //shift right by one
            }
        }
}
```

****

Note: Similarly one can write the program for the Shifting the LED in the reverse order.

Program 4.11: Exploring Bit Capabilities of the Microcontroller

This program makes use of the microcontroller's bit programming capability. Here the single port pin is used as an output. The program toggles the single port pin to give a continuous on/off pattern on pin P1.1. Input from P1.0 is simulated.

Program Source Code

****

```
#include<reg52.h>
sbit van= P1^0;              /* Pin declaration vinod = port 1 pin 0*/
sbit new = P1^1;             /* Pin declaration new = port 1 pin 1*/

void main(void)
  {
      int vinod;                      // Define the int vinod
      vinod = 0x00;
      while(1)                /* Infinite loop (i.e.,countinuos loop)*/
      {
      van = vinod;      /* Simulating input from port pin P1.0 and
                        assigning the same to character van */
      new = vinod;                /* send output to port1 pin1 */
      vinod=!vinod;               //compliment vinod
      }                           // end of loop
        }                                // End of main
```

****

4.6 More Projects on LED Interfacing to Microcontrollers

Based on the programs given above many advanced projects are possible. Given below is the list of project ideas, which can be explored with little efforts by modifying the LED interfacing circuit.

4.6.1 Running LEDs

http://www.avrprojects.net/articles.php?lng=en&pg=43

The code for the running LEDs has already been developed in this chapter. The project at the above URL describes running LED or LED chaser project. It has running LED light with 15 red 3 mm LEDs. Just have a look at the project and it is very simple to implement on our kit.

4.6.2 Running Bicolor LED

The above project can be very easily modified by connecting the readily available bicolor LED package such as TLUV5300 (Refer: www.vishay.com/leds/list/product-83056) which comprises green and red LEDs. Different patterns can be created by modifying the source code.

4.6.3 Interfacing 6 LEDs Using 3 Microcontroller Pins by

URL: http://www.scienceprog.com/connect-6-leds-using-3-microcon troller-pins/ A simple circuit is given for expanding the pin capabilities useful when using dual colour LEDs, i.e., 2 LEDs packed in one case but in different directions. AVR-GCC C code to control the LEDs is also given. With the similar scheme even 12 LEDs can be connected by using only 4 port pins.

4.6.4 The LED Dimmer Demoboard

http://www.standardics.nxp.com/support/boards/leddemo/

The LED dimmer demoboard demonstrates the capability of the I2C-bus to perform red/green/blue color LED lighting and mixing operations. Two separate control cards comprise the demoboard:

- The Keypad Control Card contains a microcontroller, a 16-key keypad/controller, and a power supply/regulator module.
- The LED Control Card contains LEDs (red, green, blue, white, RGB) and the devices used to control them.

4.6.5 Fading RGB LED

URL: http://www.avrprojects.net/articles.php?lng=en&pg=53

The project at the above URL is done with the AVR microcontroller. But the same can be implemented with the little modification in the circuit diagram covered above. An RGB LED is an LED which has three LEDs integrated in one packaging. These LEDs have the colors red, green, and blue. With these three colors one can mix a good number of color and fading combinations using the 8 bit PWM drive. With 8-bit PWM drive 256*256*256 combinations are possible, which leads to 16.777 million colors, and 256 different brightness.

With the three LEDs driven with logic 1 at port pins gives white color. The RGB LED has generally four leads, one for each color and one for the common cathode which is connected to ground. The operating voltages for every LED is different, viz., Red LED 2 V, Green 3.5 V, and the Blue LED 3.5 V with each LED drawing around 20 mA current. The timers of the microcontroller can be used to drive the particular pattern which corresponds to fading of the LEDs. You may also use IC555 output which is available on the kit.

4.6.6 LED Moving Font

URL: http://www.woe.onlinehome.de/e_projects.htm

Page title: World of Electronics – Electronic Projects

The project at the above URL describes a microcontroller (8051-based) LED moving font board built up of separate modules consisting of 64 LEDs each (8x8 matrix). There is a provision of cascading the modules according to the desired size of the font. Each module is controlled by the LED display driver MAX7219 (or MAX7221) which can drive 64 LEDs. The display data is transferred serially to this display driver via the pins DIN, CLK, and LOAD. The pin DOUT can be connected to the input DIN of the following display driver, all CLK and all LOAD pins are connected together. The modules are controlled by an 8051-compatible microcontroller AT89C51 (LED moving font controller variant 1 or AT89C2051 (LED moving font controller variant 2 from Atmel which provide 4 KB or 2KB flash memory on-chip.

4.7 DIP Switch Interfacing: Getting Input to Your Embedded System

Dual in-line package (DIP) switches are electronic packages that consist of a series of tiny switches. DIP switches are used to configure computers and peripherals such as circuit boards and modems. They are really useful for testing purpose. On our kit an eight input DIP switch is provided with all its inputs available through the male connector on the board. Any port can be connected to this DIP switch by simply connecting the associated male connectors. This scheme is very useful to simulate a digital input to the microcontroller.

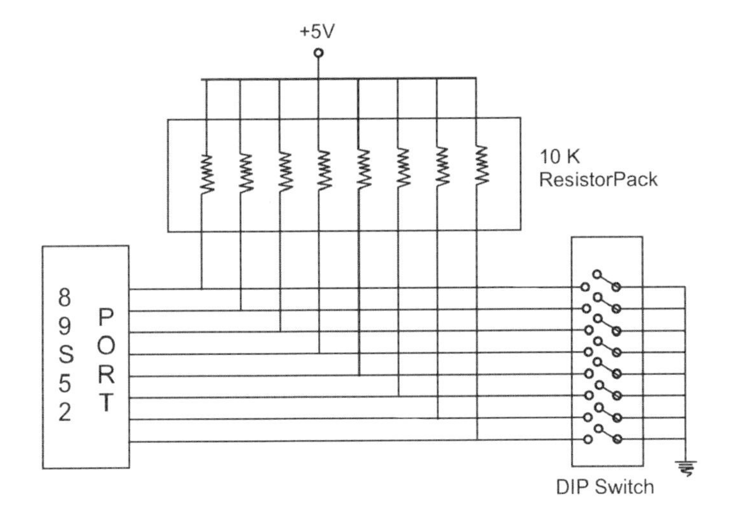

Figure 4.4 DIP switch interfacing to microcontroller

Program 4.12: DIP Switch Interfacing

Program for getting the DIP switch status and outputting the same on LED. DIP switch status is read through port 1. LEDs are connected to port 0 to display the status read.

Program Source Code

```
*****************************************************
#include <REG52.H>
void main(void)
{
 char N1;                    //Define the character for assignment of the
hex number.
        P1 = 0xFF;            // Define the port P1 as a input port.

        while(1)

        {
              N1=P1;        // Assign the value read from the port 1 to N1.
              P0=N1;         /* display the status of the DIP switch on the
                                    LEDs connected to port 0.    */
        }                   // End of "while"
}                           // End of "main"

*****************************************************
```

4.8 LCD Interfacing

LCDs are commonly used as an output interface to embedded systems products like cell phone, data loggers and process control instruments. Figure 4.5 shows LCD controller Hitachi 44780 that provides a relatively simple interface between a processor and an LCD. Since the LCD interfacing is used in almost all our applications, a datasheet and other details are given in annexure I. Three output pins, viz., P2.0, P2.1 and P2.2 are used for controlling the LCD. The data is passed on to the LCD using port 1.

The most important thing in LCD interfacing is to apply a minimum of 450 ns wide high to low pulse to the enable pin of the LCD. This is used to activate the latch of Hitachi 44780 to grab the data on the port pin at that instance.

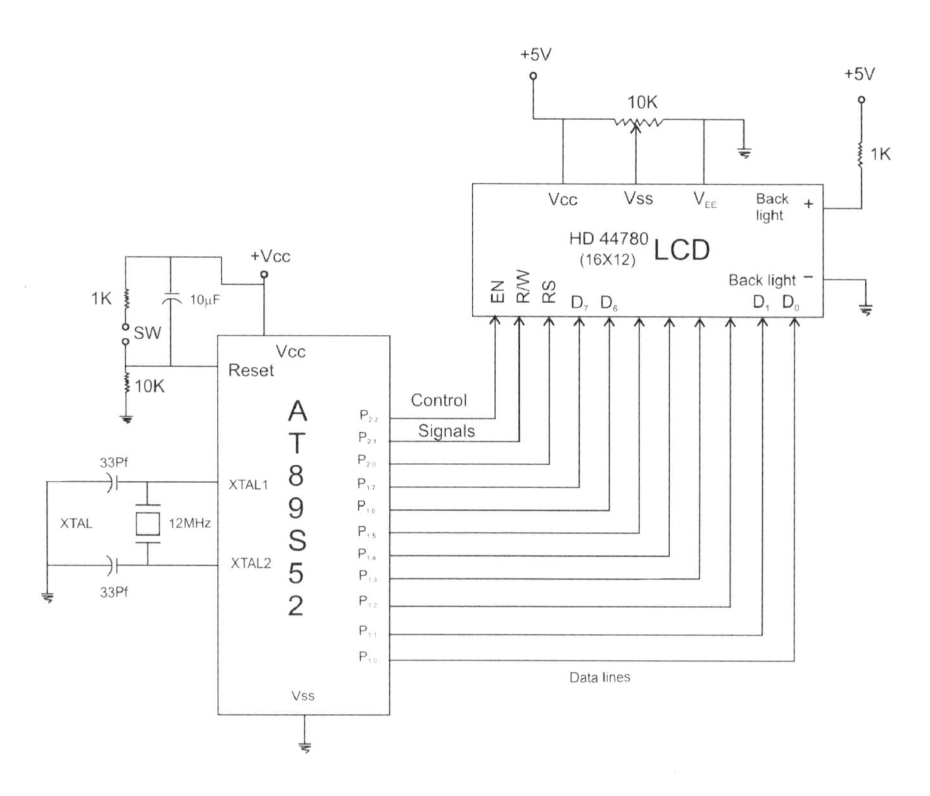

Figure 4.5 LCD interfacing to 8051 microcontroller

Program 4.13: Displaying String of Characters on LCD

This program displays the string "We Thank Springer Publishers From Authors" on the LCD. The comments are self-explanatory.

Program Source Code

```
*****************************************************
#include <REG52.H>        /* special function register declarations  */
                          /* for the intended 8051 derivative     */
sbit RS = P2^0;       // Control signal RESET of the LCD connected
                         to pin P2.0
sbit RW = P2^1;       // Write (RW) Signal pin connected to pin P2.1
sbit EN = P2^2;       // Enable (EN) LCD control signal connected to
                         pin P2.2
int i;
void delay(void);        // Delay function
void INIT(void);     // Function for the commands to Initialization of
                         the LCD
void ENABLE(void); // Function for the commands to enable the LCD
void LINE(int);  /*  Function for the commands to choose the line for
                    displaying the characters on the LCD */

void LINE(int i)     // start of LINE function
{
        if (i==1)
        {
        RS=0;     // first make the reset signal low
        RW=0;     // make RW low to give commands to the LCD
        P1=0x80; // force cursor to beginning of the first line
        ENABLE(); // enable the LCD
        RS=1;
        }
        else
        {
        RS=0;
        RW=0;
        P1=0xC0;     // force cursor to beginning of the second line
        ENABLE();
        RS=1;
        }
}
```

```
void delay() // Invoking the delay function.
{
int j;
for (j=0;j<10;j++)
{
for (i=0;i<100;i++);
}}

void ENABLE(void)            // Invoking the enable function
{ EN=1;     // Give high to low pulse to Enable the LCD.
delay();
EN=0;
delay();
}

void INIT(void)        // Initialization of the LCD by giving the proper
                          commands.
{
        RS=0;
        RW=0;
        EN=0;
        P1 = 0x38;   // 2 lines and 5*7 matrix LCD.
        ENABLE();
        ENABLE();
        ENABLE();
        ENABLE();
        P1 = 0x06;   //Shift cursor to left
        ENABLE();
        P1 = 0x0E;   // Display ON, Cursor Blinking
        ENABLE();
        P1 = 0x01;   // Clear display Screen.
        ENABLE();
}


void main (void)
{

char code dis[] = "We Thank Springer Publishers";
char code dis2[] = "From Authors";              // Array of characters
                                                   to be display.
```

```
char *p;                // Pointer to point the data in the array.
int j,x;
while(1)
{
for (x=1;x<3;x++)
{

INIT();
LINE(1);      // Display on the line 1 of the LCD.
if (x==1)
{
p=&dis;       // point the character address from the array "dis" where
                 it is stored.
}
else
{
p=&dis2;      // Point the Character address from the "dis2" where it
                 is stored/
}
for (j=0;j>8;j++){
        P1= *p++;       // Increment the pointer by 1.
        ENABLE();
        }
LINE(2);   // Display the second array on the second line on the LCD.
}
}

}
```

****

4.9 Interrupts in Microcontrollers

Interrupts are of immense importance in any microcontroller-based system. A peripheral such as timer/counter, A to D converter, or even a memory or a keyboard can interrupt the microcontroller from its ongoing sequential execution to let it know that it requires attention. This may be due to several reasons such as timer overflowing, ADC returning data, or the signal on an input pin changing. Generally, many standard references compare the polling technique with the interrupt and concludes that the interrupt is a superior technique. However, in microcontroller paradigm, if the developer intends to use the polling technique, then he will have to implement it in software. Ideally the peripheral can be made to set/reset a flag, the status of which requires to be checked using a conditional loop. But the interrupt offers a flexible and time-efficient response to the external events with the least software overhead.

Interrupts in 8051 microcontroller are as follows:

- Timer 0 Overflow
- Timer 1 Overflow
- Reception/Transmission of Serial Character
- External Event 0
- External Event 1

4.9.1 Writing ISRs in C

Program 4.14:

This simple program illustrates the initialization sequence for the timer, serial I/O and interrupt. A function called "serial_int" is set as the handler for interrupt 4, which is the serial port interrupt. A character 'V' is sent to the serial buffer. Observe this in the simulation window of Keil IDE.

Program Source Code

```
**************************************************

// Serial Port Interrupt

#include <reg52.h>
#include <stdio.h>

    void main (void)
    {
        int i;
        TMOD = (TMOD & 0xF0) | 0x20;    // Set Timer 1
                                           Mode 2
        SCON = 0x50;                    // Serial Communi-
                                           cation Mode 1
        EA = 1;                         // Enable Global
                                           Interrupt Bit
        ES = 1;                        // Enable Serial Port
                                           Interrupts
```

```
        TH1 = 0xFD;                 // Load Timer 1 Baud Rate
        TR1 = 1;                    //Set Timer 1 Run control
                                      bit


        while (1)
        {
            SBUF = 'V';             //Send V continuously on
        //Serial Port
            for(i=0;i<=10;i++);     /* Delay to avoid trans-
                                      mission overlapping */
        }
    }


// Serial Port Interrupt Service Routine.

        void timer0_ISR (void) interrupt 4
        {
            TI = 0;                         //Clear transmission
                                              interrupt flag
        }
*************************************************************
```

Program 4.15:

This program is modification of the Program 4.14. Unlike the interrupt technique used in Program 4.14, here the interrupt flag is polled continuously. Observe the transmission of 'v' in simulator window.

Program Source Code

```
*************************************************************

// Serial port Interrupt Programming

#include<REG52.H>

void main()
{

    TMOD=0x20;                              // Timer 1 Mode 2
    TH1=0XFD;                   // Baud rate = 9600 for 11.0592
```

```
MHz XTAL          TL1=0XFD;

        SCON=0X50;                          //Serial Communication
                                              mode 1

        TR1 = 1;                            // Start Timer 1
        TI = 0;                             // Clear Transmit Interrupt
                                              flag
        while(1)
              {

              SBUF = 'v';                   // Send 'v' continuously on
                                              Serial port
              while(!TI);                 // Check TI flag till it is zero
              TI = 0;                       // Clear Transmit Interrupt
                                              flag

              }
}
```

****

Program 4.16: Implementation of Minutes Counter

This program implements a seconds counter on the serial seven-segment LED. Timer 0 is used for the generation of delay of a minute. With elapsing of the delay routine the count is incremented and displayed on the serial seven-segment LED. The 4 seven-segment display is serially connected. Pins P3.7 is used for clock generation and the data is transferred serially over the pin P3.6.

Program Source Code

```
*********************************************************
#include<REG52.H>
void display(char a);
void clear(void);
void INIT (void);
void delay(void);
unsigned int count, milliseconds; // Unsigned Integers for the count.
sbit dat=P3^6; // Set the pin to Sending the data bits over the pin
                    P3.6
sbit clk=P3^7; // Set the pin to send clk
```

```
unsigned int delay1;
unsigned char s[10]= {0xc0,0xf9,0xa4,0xb0,0x99,0x92,0x82,0xd8,0x80,
                      0x90};

void delay(void) // Invoking the Delay routine
{
int c,d;
for(c=0;c<2;c++)
{for(d=0;d<2000;d++);
}}
void INIT (void) // Invoking the function to start the timer
{
TMOD=0X01;
TF0=0;      /* Timer 0 overflow flag reset first, Start the timer and
            set the interrupt when timer overflows */
TR0=0;
delay1=50000;
delay1=~delay1; // Complement the delay count
TL0=delay1%256;  // Count loaded in the Timer 0 lower byte.
TH0=delay1/256;  // Load the higher byte of timer 0
TR0=1;           // Run the timer 0
ET0=1;
EA=1;            // Global interrupt bit
IP=0x00;
}

void clear(void)     /* Invoking the function to clear the display */
{
int j;
dat=1;
for(j=0;j<32;j++)
{
clk=0; // Sending the pulse to clear the display
clk=1;clk=0;
}}

void display(char a) // Display the numbers on the seven-segment
                       LEDs
{
unsigned char mask;
int y;
mask=0x80;                // Mask the higher digit
```

```
for(y=0;y<8;y++)
{
if(a&mask)
dat=1;
else
dat=0;
clk=0;
clk=1;clk=0;
mask=mask>>1;    /* If the unit digit goes overflow then switch the
                    upper digit. */
}
}

void main(void)             // Invoking the main function
{
unsigned char m,n;
clear();
count=0; milliseconds=0;
 INIT();                    // Initializing the timer
do
  {
  clear ();
       n=count%10;
       m= ( (count/10) % 10);
  display(s[m]);        // Display the count at higher digit position
  display(s[n]);        // Display the count on the lower digit position
delay();
}
while (count < 60);
 TR0=0;
  while (1);
  }


(void) interrupt 1     // Function for the interrupt routine when timer
                         overflows
{
TR0=0; TF0=0;          // Reset the timer and the timer flag
delay1=50000;
delay1=~delay1;
TL0=delay1%256;
TH0=delay1/256;
```

```
milliseconds++;     // Increment the millisecond count by one
if (milliseconds == 20)
  {
count ++;           // Increment the count
milliseconds=0;
}
TR0=1;     // Run the timer 0
}
*****************************************************
```

4.9.2 A Word about Interrupt Latency

Interrupt latency is the time between the generation of an interrupt by a device and the servicing of the device which generated the interrupt. In other words it is the time taken to service the interrupt or the longest time between when the interrupt occurs and when the microcontroller suspends the current processing context.

Interrupt latency is an important issue in microcontroller-based system design for several reasons. Microcontroller-based system with a single interrupt will execute correctly (as per the intended logic of the developer) provided ISR execution time is smaller than the frequency of the interrupt. One of the most important issues in microcontroller-based embedded systems is the power consumption, since the system aims at remote installation with long battery life. From the low power consumption point of view, the crystal frequency should be at the lowest end of the permissible limit. This value is ultimately decided by the interrupt latency (generally the standard values should be adopted to give prescribed baud rates in serial communication). Improving the interrupt latency results into lowering down the throughput and increasing the processor utilization. On the other hand, reducing the processor utilization may increase interrupt latency but may decrease the overall throughput.

Program 4.17: Estimating Interrupt Latency

Program Source Code

```
*****************************************************

#include <REG52.H>          /* special function register declarations */
                                /* for the intended 8051 derivative */

sbit RS = P2^0;
sbit RW = P2^1;
```

```
sbit EN = P2^2;

sbit interrupt1 = P3^3;              // External interrupt

sbit IN =P2^3;
void delay(int);
void INIT(void);
void ENABLE(void);
void LINE(int);
int ONTIME(int);

void LINE(int i)     // Enabling the Line function
{
        if (i==1)
        {
        RS=0;
        RW=0;
        P1=0x80;
        ENABLE();
        RS=1;
        }
        else
        {
        RS=0;
        RW=0;
        P1=0xC0;
        ENABLE();
        RS=1;
        }
}

void delay()
{
int i,j;
for (j=0;j<10;j++){
for (i=0;i<100;i++);
}}

void ENABLE(void)      // Invoking the Enable function for the LCD.
{
EN=1;
delay();
```

```
EN=0;
delay();
}
void INIT(void)   // Initialization of the LCD by giving the proper
                     commands
{
      RS=0;
      RW=0;
      EN=0;
      P1 = 0x38;      // 2 lines and 5*7 matrix LCD
      ENABLE();
      ENABLE();
      ENABLE();
      ENABLE();
      P1 = 0x06;      // Shift cursor to left
      ENABLE();
      P1 = 0x0E;      // Display ON, Cursor Blinking
      ENABLE();
      P1 = 0x01;      // Clear display Screen
      ENABLE();
}

int ONTIME()                     // Invoking the ONTIME function
{

TMOD=0x01;          // Intialize as 16 bit timer(mode 1)(Timer 1)
TL0=0x00;
TH0=0x00;
TR0=1;          // start the timer
IN=1;           /*input to 8052 from 555 osc*/
while(IN);
while(!IN);
while(IN);
TR0=0;        // stop the timer


return((TH0*256)+TL0); // return the ISR Execution count
}


void main (void){
int latency, unit, tens, hundred, thousand,
```

```
tenthou, unit1, tens1, hundred1, thousand1, tenthou1;
char code dis[] ="ISR Execution Time=";
char code dis2[] ="Interupt Latency=";
char *p;
int i,j;

while(1)
{
TMOD=0x01;      // Initialize the timer to count the total program
                   execution time
TH1=0x00;
TL1=0x00;
TR1=1;             // Start the timer 1
if(interrupt1==1)
{
i = ONTIME();
unit = (i%10);
tens = (i/10)%10;
hundred = (i/100)%10;
thousand = (i/1000)%10;
tenthou = (i/10000)%10;
unit = unit + 0x30;          // Sending the ASCII data to the LCD
tens = tens + 0x30;
hundred = hundred + 0x30;
thousand= thousand + 0x30;
tenthou = tenthou + 0x30;

INIT();
LINE(1);

p=&dis;
for (j=0;j<8;j++){
      P1= *p++;
      ENABLE();
      }
LINE(2);
P1=tenthou;
ENABLE();
P1=thousand;
ENABLE();
P1=hundred;
ENABLE();
```

```
P1=tens;
ENABLE();
P1=unit;
ENABLE();
}
else
{
p=&dis2;
for (j=0;j<8;j++){
      P1= *p++;      // Pointer to point the data from the array
      ENABLE();      // Enabling the LCD
      }
LINE(2);     // Display the content on the second line of the LCD
TR1=0;                              // stop the timer 2
latency=(((TH1-TH0)*256)-(TL1-TL0)); //Timer 0 count is subtra-
                                            cted from timer1

unit1 = (latency%10);
tens1 = (latency/10)%10;
hundred1 = (latency/100)%10;
thousand1 = (latency/1000)%10;
tenthou1 = (latency/10000)%10;
unit1 = unit1 + 0x30;
tens1 = tens1 + 0x30;
hundred1 = hundred1 + 0x30;
thousand1= thousand1 + 0x30;
tenthou1 = tenthou1 + 0x30;
LINE(2);
P1=tenthou;                         // display the latency
ENABLE();
P1=thousand;
ENABLE();
P1=hundred;
ENABLE();
P1=tens;
ENABLE();
P1=unit;
ENABLE();
}
}
}
*******************************************************
```

Chapter 5

Hyperterminal-Based Control

5.1 Hyperterminal

HyperTerminal is a terminal emulator program that runs in MS Windows. It offers a text-based command prompt interface on a remote device. The remote device is a serial device, and it could be a router, connected directly to the PC's serial port OR a network device. HyperTerminal is generally used for the local serial interface for communications or the network. In the case of the network, HyperTerminal is simply using the telnet protocol. In order to communicate with the device, the HyperTerminal must be configured on the right COM port (either COM1 or COM2) and at appropriate baud rate (most of the time 9600 for applications written here).

Having designed to emulate various types of text terminal configurable to make a over a serial port the serial parameters to be selected are:

- Baud rate
- Parity (odd, even, none)
- Stop bits
- Flow control

5.2 Packet-based Control Scheme

The beauty of the programs covered in this section is that they present an insight with regard to the new protocol development. The remote host, i.e., microcontroller board connected to COM1, has various command blocks which will be invoked based on the parameters passed from the HyperTerminal. The command blocks are packed and a particular packet corresponding to the desired control sequence that is made alive as soon as the associated command parameter is received from the HyperTerminal.

J.S. Parab et al. (eds.), Exploring C for Microcontrollers, 69–83.

Two things are required to keep in mind in these applications: maintaining the compatibility of PC serial port to microcontroller; and selection of proper baud rate. A standard serial interfacing for PC, RS232C, requires negative logic, i.e., logic '1' is −3 V to −12 V and logic '0' is +3 V to +12 V. In order to convert a TTL logic, say, TxD and RxD pins of the microcontroller chips, a converter chip is required. A MAX232 chip is a very popular chip used in the circuit boards for all the applications covered in this chapter. It provides 2-channel RS232C port and requires external 10uF capacitors. The polarity of the capacitor is crucial which the user must note. Another possibility in lieu of MAX232 is DS275, which does not require an external capacitor.

5.3 Mechanism and Lots of Possibilities

The mechanism of packet-based interface works as follows. The number of control possibilities are listed in case of each control module and categorized in terms of packets. For instance, in case of the LED interface board, various display sequences are defined as packets. In case of stepper motor-based interface the packet definitions include the clockwise, anticlockwise rotations in the first instance and step size as the secondary instance, both the things defined together. As per the firing of the commands from the HyperTerminal the firmware residing in the microcontroller flash will invoke the appropriate control action.

The following simple steps will help you to configure the HyperTerminal.

1. Open the HyperTerminal, by clicking Start>Programs>Accessories>Communications>HyperTerminal.

2. This will prompt you to create a new connection. The new connection can also be created by clicking File>New Connection. Enter a name for the connection and click OK.

3. Select the serial port where the device is connected to the PC (COM1 in our case). Click OK.

4. Enter the following configuration for the connection:

Bits per second	9600
Data bits	8
Parity	None
Stop bit	1
Flow control	None

Click OK.

5. After clicking OK or apply all, the above configuration will be loaded and the commands can be entered at the command prompt.

This HyperTerminal-based interface offers great possibilities from control point of view. With little modification in the hardware it is possible to go for web enabling of the embedded products by adding an Ethernet-based interface. Internet-based data logging and control, microcontroller-based server, and client applications are possible with optimized usage of resources due to packet methodology.

5.4 Application 1: Packet-based Interface for LEDs

The circuit diagram of the packet-based LED interface is as shown in Figure 5.1. The commands to be emulated from the HyperTerminal and the corresponding control action is given in the program source code.

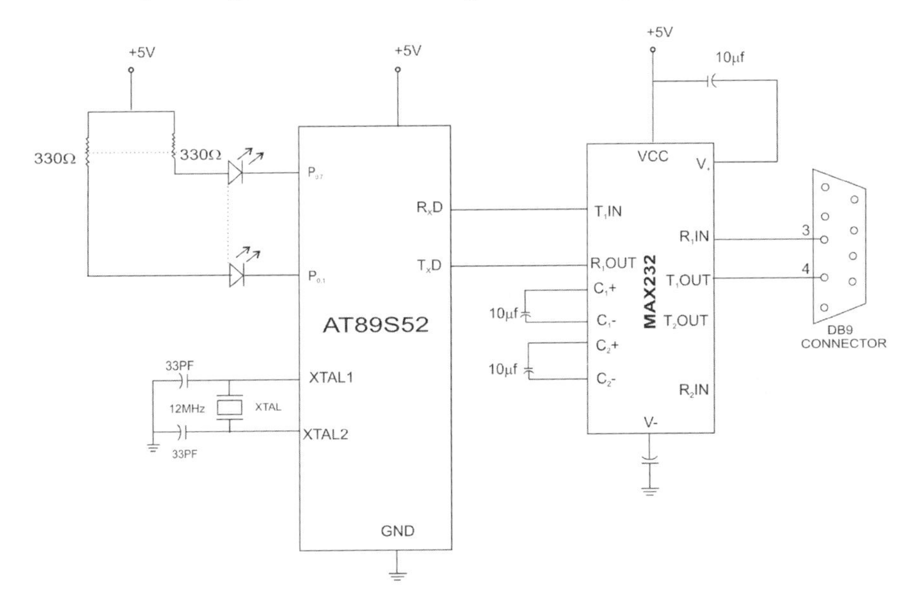

Figure 5.1 Circuit diagram of packet-based interface for LED

Program Source Code

```
*******************************************************

/* Displaying different LED sequence pattern based on the packet
received */
#include<REG52.h>
```

```
unsigned char led;
int interpret;
char packet[4];
char index;
void init(void);
void delay (int m);

void init()
{
TMOD=0X20;
TH1=0xFD;                    /*select baud rate 9600*/
SCON=0x50;
TR1=1;                       /* start timer*/
}

void main(void)
{
int k=0;
char i;
init();
ES=1;
EA=1;          // Global interrupt bit enabled
led=0xff;
for (i=0;i<4;i++){
packet[i]=0;
}

index=0;
interpret=0;

while(1)
{
    {
     P0= led;
     delay(1000);
    }

    }}
void delay (int m) // Function for the Delay.
{
     int z;
    z=m;
```

```
    while (m>0)
    { m--;}
}

void serialisr() interrupt 4
{
    unsigned char y;
    char i;
    RI=0;
    y=SBUF;      // write the data in srial buffer.
    TI=0;
    SBUF=y;      // read the data from serial buffer.
    while (!TI); // Check the flag
    TI=0;
    packet[index]=y;
    index++;
    if (y=='z')
        {
        interpret =1;
        index = 0;
        }
        else interpret = 0;
        if (interpret==1)
        {
        switch(packet[0])
        {
        case 'A':
                led=0x55;    // 01010101
                break;

        case 'B':
                led=0x33;
                break;

        case 'C':
                led=0x55;
                delay(100);
                led=0x33;
                break;

        case 'D':
                led=0x00;
```

```
                delay(100);
                 led=0x02;
                 delay(100);
                led=8;
                delay(100);
                 led=16;
                 delay(100);
                 led=16;
                delay(100);
                led=32;
                delay(100);
                led=64;
                delay(100);
                led=256;
                delay(100);
                break;

        }

    for (i=0;i<4;i++)
    {
        packet[i]=0;
    }
    interpret=0;
    }
}
```

```
**************************************************
```

5.5 Application 2: Packet-based Interface for Stepper Motor Control

The circuit diagram of the packet-based stepper motor interface is as shown in Figure 5.2. The commands to be emulated from the HyperTerminal and the corresponding control action are given in the program source code. The interface is based on UN2003 driver popularly used for many stepper motor control applications. It is a 7-bit, 50 V 500 mA TTL-input NPN Darlington driver. The drivers does not need power supply; the VDD pin is the common cathode of the seven integrated protection diodes. The details are covered in may web-based resources [35].

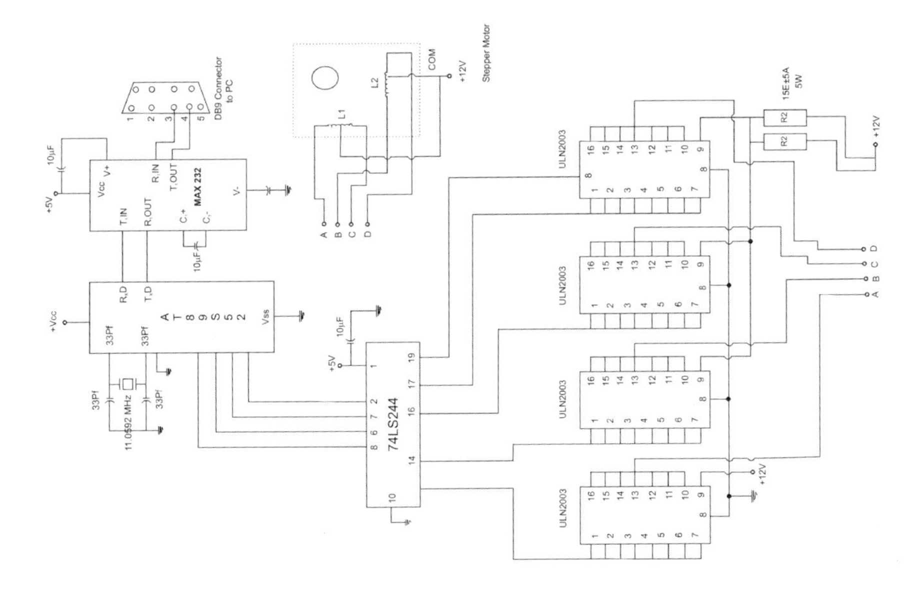

Figure 5.2 Circuit diagram of stepper motor interface

Program Source Code

```
*************************************************

#include<REG52.h>
unsigned char mot[4] = {0x0a,0x06,0x05,0x09};
unsigned char run, clkwise;
int interpret, mode, count;
char packet[5];    /*p[0] is control char*/
char index;    /*others may contain data last must contain end char,
               i.e., 'z' */
void init(void);
void delay (int m);
void init()
{
TMOD=0x20;
TH1=0xFD;    /*select baud rate 9600*/
SCON=0x50;
TR1=1;    /* start timer*/
}
```

```
void main(void)
{
int k=0;
 char i;
 init(); ES=1; EA=1;
run=1;
clkwise = 1;
for (i=0;i<5;i++){
packet[i]=0;
}
index=0;
interpret=0;
while(1)
{
while(mode==0){    /*continues rotation*/
while(run==1)
{if(clkwise==1){
  for (i=0;i<4 ; i++)    /* clkwise dir*/
    { P0= mot [i];
     delay(1000);
    }}
          else
          {for (i=3;i>=0 ; i--) /*anticlkwise dir*/
    { P0= mot [i];
     delay(1000);
    }
          }
          }
}
}

while(mode==1) /*number of step's given by user*/
    {
    while((run==1)&&(count>0))
          {
          if(clkwise==1){
          for ( i=0;i<4 ; i++)
          {
                    P0= mot [i];
                    delay(1000);
                    }
          }
```

```
else
        {
        for ( i=3;i>=0 ; i--)
                {
                P0= mot [i];
        delay(1000);
         }
         }
    count--;
    }
  }
}
  void delay (int m)
  { int z;
  z=m;
   while (m>0)
   { m--;}
  }

void serialisr() interrupt 4
{
  unsigned char y;
  char i;
  RI=0;
  y=SBUF;                                    /* contents of buffer
                                             in y*/
  TI=0; SBUF=y; while (!TI);TI=0;            /*similtaneusly
                                             display*/
  packet[index]=y;
index++;
if (y=='Z') {
  interpret =1; /* indicates end of packet*/
    index = 0;
          }
  else interpret = 0;
  if (interpret==1)
    {
    switch(packet[0])
    {
    case 'S':
          run=0;
          break;
```

```
case 'R':
        run=1;
        break;

case 'A':
        clkwise=0;
break;

case 'C':
        clkwise=1;
        break;

case 'D':
        if (packet[1]=='0')
        mode=0;
        break;
case 'E':
        if (packet[1]=='1')
        mode=1;
        break;
case 'F': /*variable size packet*/
        count=(packet[3]-0x30)+((packet[2]-0x30)*10)+((packet[1]-
0x30)*100);
        break; /*packet[3]=units..........*/
        }
for (i=0;i<5;i++){
packet[i]=0;
}
interpret=0;
}
}
*****************************************************
```

5.6 Application 3: Home Automation from PC HyperTerminal

Home Automation from PC Terminal

With the deep penetration of PCs in today's modern homes, the trend of PC-based home automation is growing. There are different home automation systems that consist of controller and device controlling modules (relays, light dimmers, etc.). New protocols such as X10 system that communicates through mains wiring is gaining popularity.

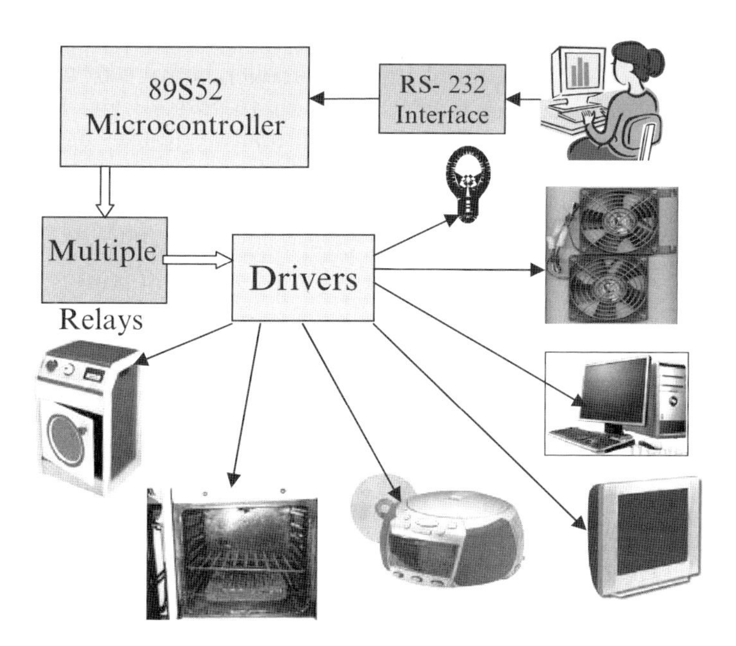

Figure 5.3 Block schematic of the hyper terminal based home automation system

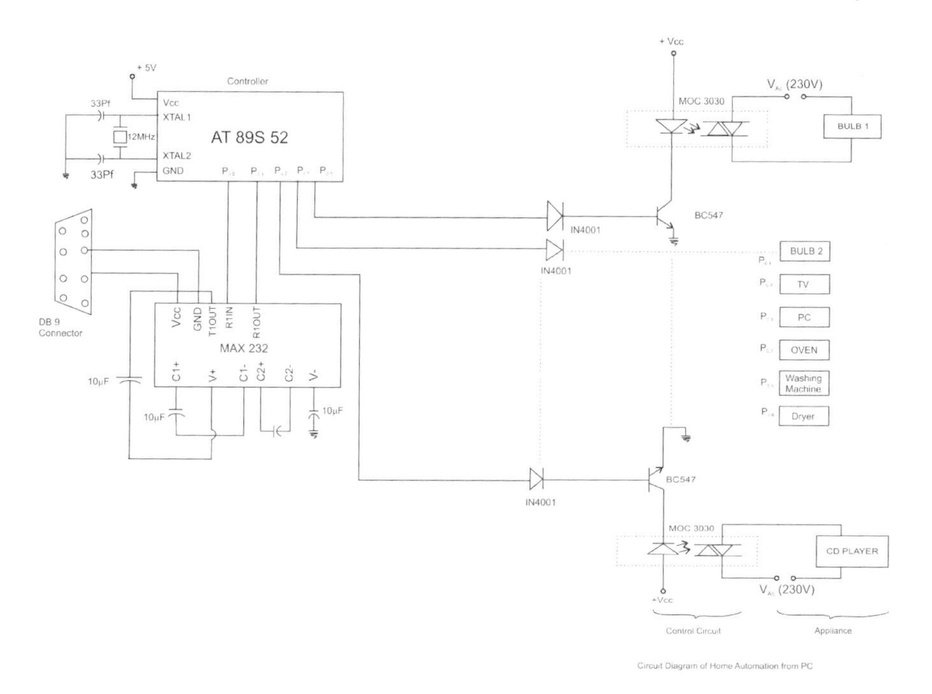

Figure 5.4 Circuit diagram

However, any embedded developer would definitely love to write his own protocol for the home automation system. Here is the one which communicates with the 89S52-based microcontroller board through HyperTerminal of PC. Again the packet-based methodology has been used. The drivers for various appliance control such as lights, garage door, audio system, TV, etc., resides in the microcontroller flash and are invoked based on the command character passed on from the HyperTerminal. The microcontroller-based interfacing board comprises optoisolator MOC3030 which has built-in zero crossing detection.

Program Source Code

```
*************************************************************


#include<REG52.h>
unsigned char light;
int interpret;
/* The pins of the port 0 are connected to power line of various
home appliances through the control circuitry to control them by
on/off the power supply. */

sbit light1=P0^0;    // Pin connected to the control circuit of the
                        Light 1.
sbit light2=P0^1;
sbit TV=P0^2;    // Pin connected to control circuit of power supply
                   of TV.

sbit PC = P0^3;
sbit oven = P0^4;
sbit washing = P0^5;
sbit coller = P0^6;
sbit CD = P0^7;

char packet[4];
char index;
void init(void);
void delay (int m);
void init()
{
TMOD=0X20;
TH1=0xFD;              /*select baud rate 9600*/
SCON=0x50;
```

```
TR1=1;                  /* start timer*/
}

void main (void)
{
int k=0;
char i;
init(); ES=1; EA=1;
for (i=0;i<4;i++){
packet[i]=0;
}
index=0;
interpret=0;

while(1)
{
    {
     P0=light1;
     delay(1000);
    }

    }}

void delay (int m)
{ int z;
z=m;
 while ( m >0)
 { m--;}
}

void serialisr() interrupt 4

 {
unsigned char y;
char i;
 RI=0;
 y=SBUF;
 TI=0; SBUF=y; while (!TI);TI=0;
 packet[index]=y;
 index++;
 if (y=='z') {interpret =1; index = 0;}
 else interpret = 0;
```

```
if (interpret==1)
{
switch (packet[0])
{
case 'A':
light1=1;          // switch on the bulb1
break;

case 'B':
light1=0;          // switch off the bulb1
break;

case 'C':
light2=1;          // switch on the bulb2
break;

case 'D':
 light2=0;         // switch off the bulb2// // switch on
Washing machine
break;
case 'E':
 TV=1;             // switch on the TV
break;
case 'F':
TV=0;              // switch off The TV
break;
case 'G':
 PC=1;                        // switch on PC
break;
case 'H':
 PC=0;                        // switch off the PC
break;
 case 'I':
oven=1;                       // switch on oven
break;
 case 'J':
oven=0;                       // switch off oven
     break;
case 'K':
washing=1;          // switch on the washing machine
     break;
case 'L':
```

```
washing=0;                  // switch off the washing machine
            break;
      case 'M':
coller=1;                   // switch on cooler
            break;
      case 'N':
      coller=0;                 // switch off the cooler
            break;
      case 'O':
CD=1;                       // switch on CD player
            break;
      case 'P':
CD=0;                       // switch off CD player
            break;
}

      for (i=0;i<8;i++){
packet[i]=0;
}
interpret=0;
}}
```

****

Chapter 6

Embedding Microcontroller in Routine Applications

This chapter illustrates how embedding microcontroller in day-to-day applications altogether changes the throughput of the product.

6.1 Application 1: Podium Timer

The podium timer is designed for giving the indication of timeout for a series of timed speeches. The 4×4 hex keyboard enables the user to set the time and start the timer. Once the timer is ON, it constantly indicates the time on the LCD. Before 3 minutes (which is also programmable) of actual timeout, it gives an indication by means of short beep. As soon as the time set is elapsed, the unit enables the buzzer indicating the timeout.

Program Source Code

```
***********************************************************

#include <REG52.H>          /* special function register declarations */
                            /* for the intended 8051 derivative    */
sbit RS = P2^0;
sbit RW = P2^1;
sbit EL = P2^2;
  sbit buzz = P2^3;
sbit R1 = P3^0;
sbit R2 = P3^1;
sbit R3 = P3^2;
sbit R4 = P3^3;
sbit C1 = P3^4;
sbit C2 = P3^5;
sbit C3 = P3^6;
```


J.S. Parab et al. (eds.), Exploring C for Microcontrollers, 85–122.

```
sbit C4 = P3^7;
void Delay(int);
void INIT(void);
void ENABLE(void);
void LINE(int);
int keyb(void);
void settime(void);
void starttimer(void);
void check_timeout(int);
int settime1=10;

void settime(void)
{

int ten_pos=0, unit_pos=0;
while(!(ten_pos=keyb()));
while(!(unit_pos=keyb()));
P1=(((ten_pos))+0x30);
ENABLE();

P1=((unit_pos)+0x30);
ENABLE();
Delay(2000);

settime1= ten_pos*10 + unit_pos;
}

void starttimer()
{
int set_temp= settime;

check_timeout(3); //check for the warning timeout default 3 min before
                          end time

check_timeout(set_temp);

set_temp--;
P1=(((set_temp)/10)+0x30);
ENABLE();

P1=((set_temp/10)%10+0x30);
ENABLE();
```

```
Delay(2000);

}

void check_timeout(g)
{
if(g == 3)
{
buzz=1;
Delay(100);
buzz=0;
}
if (g == 0)
buzz=1;
Delay(400);
buzz=0;
}
void main(void)
{
char test[]="Podium timer";
char code set_msg[]="1-> set time";
char code start_msg[]="2-> start timer";
char *p;
int j;

INIT();
LINE(1);
p=&test;
for(j=0;j<17;j++)
{
if(j==0)LINE(1);
if(j==8)LINE(2);
P1= *p++;
ENABLE();
Delay(200);
}
while(1){

RS = 1;
p=&set_msg;
for(j=0;j<17;j++)
{
```

```
if(j==0)LINE(1);
if(j==8)LINE(2);
P1= *p++;
ENABLE();
Delay(200);
}

RS = 1;
p=&start_msg;
for(j=0;j<17;j++)
{
if(j==0)LINE(1);
if(j==8)LINE(2);
P1= *p++;
ENABLE();
Delay(200);
}

j=0;
j=keyb();
if(j>0);

  if(j==1){
  settime();
}
if(j==2){

starttimer();
}
}
}
void Delay(int k)
{
int i,j;
for (j=0;j<k+1;j++){
for (i=0;i<100;i++);
}}

void ENABLE(void)
{
EL=1;
Delay(1);
```

```
EL=0;
Delay(1);
}
void LINE(int i){
        if (i==1) {
        RS=0;
        RW=0;
        P1=0x80;
        ENABLE();
        RS=1;
        }
        else
        {
        RS=0;
        RW=0;
        P1=0xC0;
        ENABLE();
        RS=1;
        }
}
void INIT(void)         // Initialization of the LCD by giving the proper
                          commands.
{
        RS=0;
        RW=0;
        EL=0;
        P1 = 0x38;      // 2 lines and 5*7 matrix LCD.
        ENABLE();
        ENABLE();
        ENABLE();
        ENABLE();
        P1 = 0x06;      //Shift cursor to left
        ENABLE();
        P1 = 0x0E;      // Display ON, Cursor Blinking
        ENABLE();
        P1 = 0x01;      // Clear display Screen.
        ENABLE();
}

int keyb(void){
int key=0;
C1=1;
```

```
C2=1;C3=1;C4=1;R1=0;R2=1;R3=1;R4=1;
if (C1==0) key = 1;
if (C2==0) key = 2;
if (C3==0) key = 3;
if (C4==0) key = 4;
return(key);
}
```

****

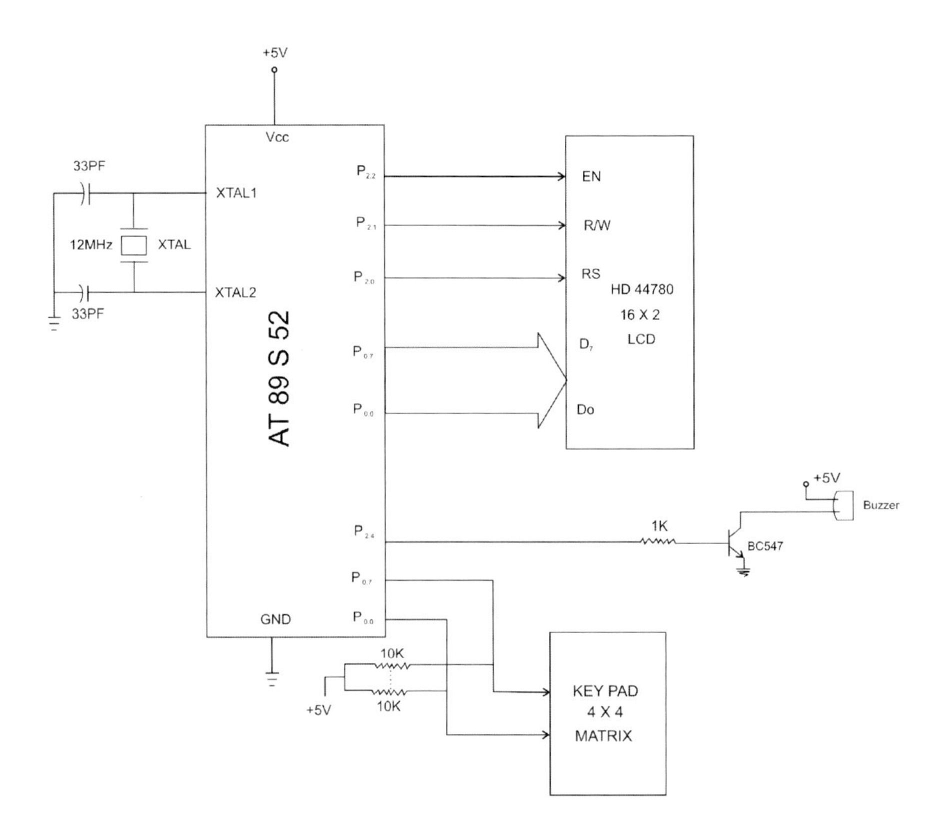

Figure 6.1 Circuit diagram of the podium timer

6.2 Application 2: Front Desk Notifier

The project aims to build an automated system that will alert the food assistance staff that a customer has arrived for service. It can also be used for a housewife who wants to know if someone has arrived at the gate. An IR transceiver pair has been used for detecting the presence of the person at the front desk. The program polls the port pins and

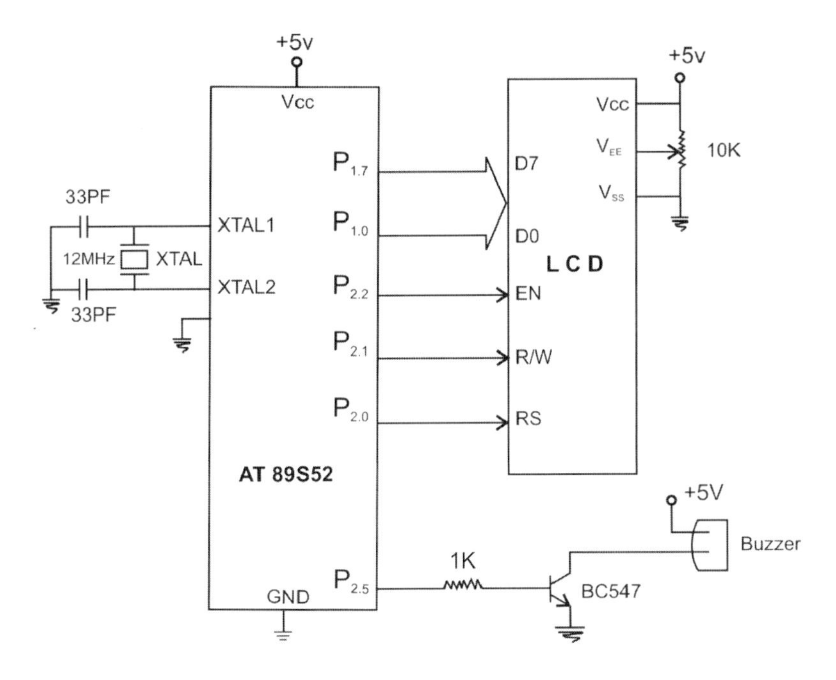

Figure 6.2 Circuit diagram of front desk notifier system

displays a message on the LCD that "Someone is arrived" as well as activates a buzzer. The system is also ideal for a bank cashier.

Program Source Code

```
*************************************************************

#include <REG52.H>   /*special function register declarations*/
                     /* for the intended 8051 derivative */

sbit RS = P2^0;
sbit RW = P2^1;
sbit EN = P2^2;
sbit IN= P2^4;       /*input from the sensor present at gate */
sbit Buzz=P2^5;      // output indication to housewife*/

void delay(int);            /* Stop Exection with Serial Intr. */
void INIT(void);
void ENABLE(void);
void LINE(int);
```

```
void LINE(int i){
        if (i==1) {
        RS=0;
        RW=0;
        P1=0x80;
        ENABLE();
        RS=1;
        }
        else
        {
        RS=0;
        RW=0;
        P1=0xC0;
        ENABLE();
        RS=1;
        }
}

void delay()
{
int i,j;
for (j=0;j<10;j++){
for (i=0;i<100;i++);
}}



void ENABLE(void)
{
EN=1;
delay();
EN=0;
delay();
}

void INIT(void)     // Initialization of the LCD by giving the proper
                       commands.
{
        RS=0;
        RW=0;
        EN=0;
        P1 = 0x38;  // 2 lines and 5*7 matrix LCD.
```

```
        ENABLE();
        ENABLE();
        ENABLE();
        ENABLE();
        P1 = 0x06;  //Shift cursor to left
        ENABLE();
        P1 = 0x0E;  // Display ON, Cursor Blinking
        ENABLE();
        P1 = 0x01;  // Clear display Screen.
        ENABLE();
}

void main (void){
char code array[] = "Some is there on the gate";
char code array1[]="No one is there on the gate";
int b,j;
Buzz=0;
while(1)
INIT();
LINE (1);
 {
if( IN==1)                         // some one is there if in = 1
{
     Buzz=1;                       // switch on the buzzer to indicate some
                                      one is there on gate
        for (b=0;b<30;b++)
                {
                 if (b==8)
                       LINE(2);
                       P1=array[b];
                       ENABLE();
                }


RS=0;
P1=0x01;
ENABLE();
RS=1;
LINE(1);
for (b=16;b<33;b++)
        {
    if (b==24)
```

```
        LINE(2);
        P1=array1[b];
        ENABLE();
        }
}
else                                    // display no one there on gate
{
for (b=0;b<30;b++)
                {
                 if (b==8)
                      LINE(2);
                      P1=array1[b];
                      ENABLE();
                }
RS=0;
P1=0x01;
ENABLE();
RS=1;
LINE(1);
for (b=16;b<33;b++)
        {
    if (b==24)
        LINE(2);
        P1=array1[b];
        ENABLE();
}
}
}
}
```

```
*****************************************************
```

6.3 Application 3: Cafeteria Food Alert/Microcontroller-based Menu Card

This economic unit is an ideal solution for small hotels. The customer can place an order of the item by pressing a key corresponding to the item he or she would like to order. The other half of the unit is the LCD display which is kept in the kitchen. The item will be displayed on the LCD and subsequently served to the customer. An interrupt will be invoked in case the customer wants bill or to place an item which is not there in menu or in case of any other events.

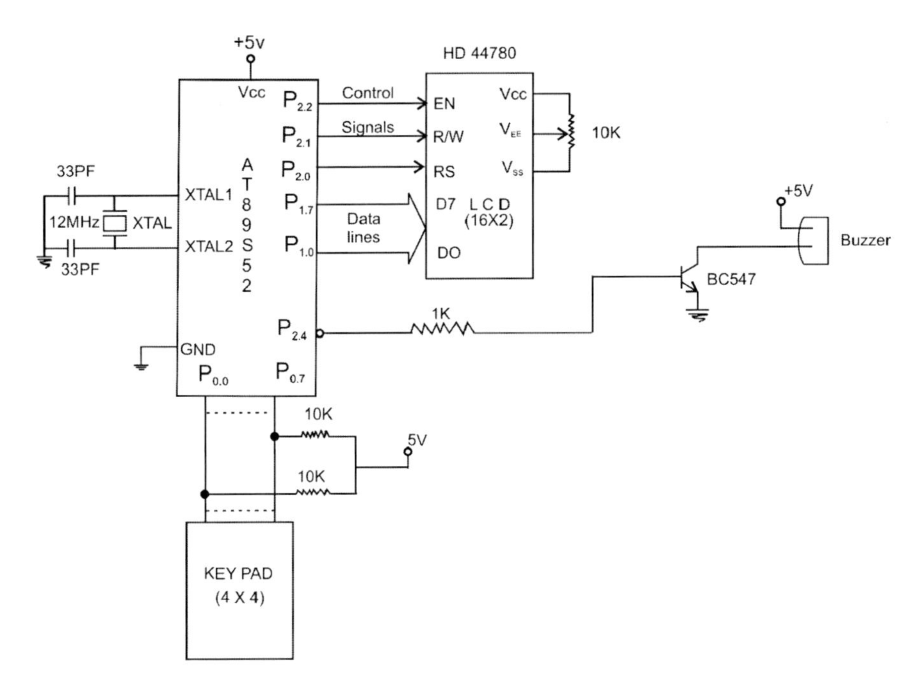

Figure 6.3 Circuit diagram of microcontroller-based menu card

Program Source Code

```
*******************************************************

#include <REG52.H>        /* special function register declarations */
#include <stdio.h>
/* for the intended 8051 derivative         */

sbit RS = P2^0;
sbit RW = P2^1;
sbit EL = P2^2;
sbit BU = P2^4;

sbit R1 = P0^0;
sbit R2 = P0^1;
sbit R3 = P0^2;
sbit R4 = P0^3;

sbit C1 = P0^4;
sbit C2 = P0^5;
```

```
sbit C3 = P0^6;
sbit C4 = P0^7;

void delay(int);              /* Stop Execution with Serial Intr. */
void INIT(void);
void ENABLE(void);
void LINE(int);
int keyb(void);

void LINE(int i){
        if (i==1) {
        RS=0;
        RW=0;
        P1=0x80;
        ENABLE();
        RS=1;
        }
        else
        {
        RS=0;
        RW=0;
        P1=0xC0;
        ENABLE();
        RS=1;
        }
}

void delay(int k)
{
int i,j;
for (j=0;j<k+1;j++){
for (i=0;i<100;i++);
}}

void ENABLE(void)
{
EL=1;
delay(1);
EL=0;
delay(1);
}
```

```
void INIT(void) // Initialization of the LCD by giving the proper
                   commands.
{
        RS=0;
        RW=0;
        EL=0;
        P1 = 0x38;      // 2 lines and 5*7 matrix LCD.
        ENABLE();
        ENABLE();
        ENABLE();
        ENABLE();
        P1 = 0x06;      //Shift cursor to left
        ENABLE();
        P1 = 0x0E;      // Display ON, Cursor Blinking
        ENABLE();
        P1 = 0x01;      // Clear display Screen.
        ENABLE();
}
int keyb(void){

int key=0;

C1=1;
C2=1;
C3=1;
C4=1;

R1=0;
R2=1;
R3=1;
R4=1;

if (C1==0) key = 1;
if (C2==0) key = 2;
if (C3==0) key = 3;
if (C4==0) key = 4;

R1=1;
R2=0;
R3=1;
R4=1;
```

```
if (C1==0) key = 5;
if (C2==0) key = 6;
if (C3==0) key = 7;
if (C4==0) key = 8;

R1=1;
R2=1;
R3=0;
R4=1;

if (C1==0) key = 9;
if (C2==0) key = 10;
if (C3==0) key = 11;
if (C4==0) key = 12;

R1=1;
R2=1;
R3=1;
R4=0;

if (C1==0) key = 13;
if (C2==0) key = 14;
if (C3==0) key = 15;
if (C4==0) key = 16;
return(key);
}

void main(void) {
char array1[]="tea";
char array2[]="wheat";
char array3[]="rice";
char array4[]="dal";
char array5[]="jaggery";
char array6[]="coffee";
char array7[]="item is not there in menu";
char array8[]="Get the bill";
char *p;
int j,i;
INIT();
LINE(1);
j=0;
```

```
while(1){
j=0;
j = keyb();
if(j==1){
for (i=0;i<8;i++){
p=&array1;
P1= *p++;
ENABLE();

}}
if(j==2){
for (i=0;i<8;i++){
p=&array2;
P1= *p++;
ENABLE();
}}
if(j==3){
for (i=0;i<8;i++){
 p=&array3;
P1= *p++;
ENABLE();
}}
if(j==4){
for (i=0;i<8;i++){
 p=&array4;
P1= *p++;
ENABLE();
}}
if(j==5){
for (i=0;i<8;i++){
 p=&array5;
P1= *p++;
ENABLE();
}}
if(j==6){
for (i=0;i<8;i++){
 p=&array6;
P1= *p++;
ENABLE();

}}
```

```
if(j==7){
for (i=0;i<8;i++){
 p=&array7;
P1= *p++;
ENABLE();

}}
if(j==8){
for (i=0;i<8;i++){
 p=&array8;
P1= *p++;
ENABLE();

}}
delay(400);
ENABLE();
delay(400);

}
}
*****************************************************
```

6.4 Application 4: Chimney Sentinel

The reports of incidents of fire due to wood-burning appliances are reported every year. This causes lots of damage to the human being as well as property. The device consists of a probe comprising a temperature sensor LM35 inserted in the chimney. This is in turn interfaced to ADC0808 and AT89S52 microcontroller-based monitoring unit. When the temperature rises the preset value set by the user, a fire is detected, and the unit sounds an alarm. There is also a provision to display the temperature on the LCD.

Program Source Code

```
*****************************************************

#include <REG52.H>   /* special function register declarations */
#include <stdio.h>
/* for the intended 8051 derivative  */
sbit RS = P2^0;
sbit RW = P2^1;
sbit EL = P2^2;
```

```
sbit BUZZ=P2^3;
sbit SOC= P2^4;

sbit a =P2^5;
sbit b = P2^6;
sbit c =P2^7;

void delay(int);
void INIT(void);
void ENABLE(void);
void LINE(int);

void LINE(int i){
if(i==1){
        RS=0;
        RW=0;
        P0=0x80;
        ENABLE();
        RS=1;
        }
        else
        {
        RS=0;
        RW=0;
        P0=0xC0;
        ENABLE();
        RS=1;
        }
}

void delay(int k)
{
int i,j;
for (j=0;j<k+1;j++)
{
for (i=0;i<10000;i++);
}
}

void ENABLE(void)
{
EL=1;
```

```
delay(1);
EL=0;
delay(1);
}

void INIT(void)       // Initialization of the LCD by giving the proper
                         commands.
{
        RS=0;
        RW=0;
        EL=0;
        P1 = 0x38;   // 2 lines and 5*7 matrix LCD.
        ENABLE();
        ENABLE();
        ENABLE();
        ENABLE();
        P1 = 0x06;   //Shift cursor to left
        ENABLE();
        P1 = 0x0E;   // Display ON, Cursor Blinking
        ENABLE();
        P1 = 0x01;   // Clear display Screen
        ENABLE();
}

void main (void)
{
int unit, tens, hundred, unit1, tens1, hundred1;
unsigned char d;
a=0;
b=0;
c=0;
while(1)
{

P1=0xff;
SOC=1;
delay(8);
SOC=0;
delay(8);
```

```
SOC=1;
delay(8);

d=P1;
if(d>40)
{
BUZZ=1;              // switch on the buzzer if temp is greater than 40
}
else
{
BUZZ=0;
}
unit = (d%10);
tens = (d/10)%10;
hundred = (d/100);

unit = (unit + 0x30);
unit1=unit+unit;
tens = (tens + 0x30);
tens1=tens+tens;
hundred = (hundred + 0x30);
hundred1=hundred+hundred;

INIT();
LINE(1);

P0=hundred1;
ENABLE();
P0=tens1;
ENABLE();
P0='.';
ENABLE();
P0=unit1;
ENABLE();
P0='c';
ENABLE();

}
}

************************************************************
```

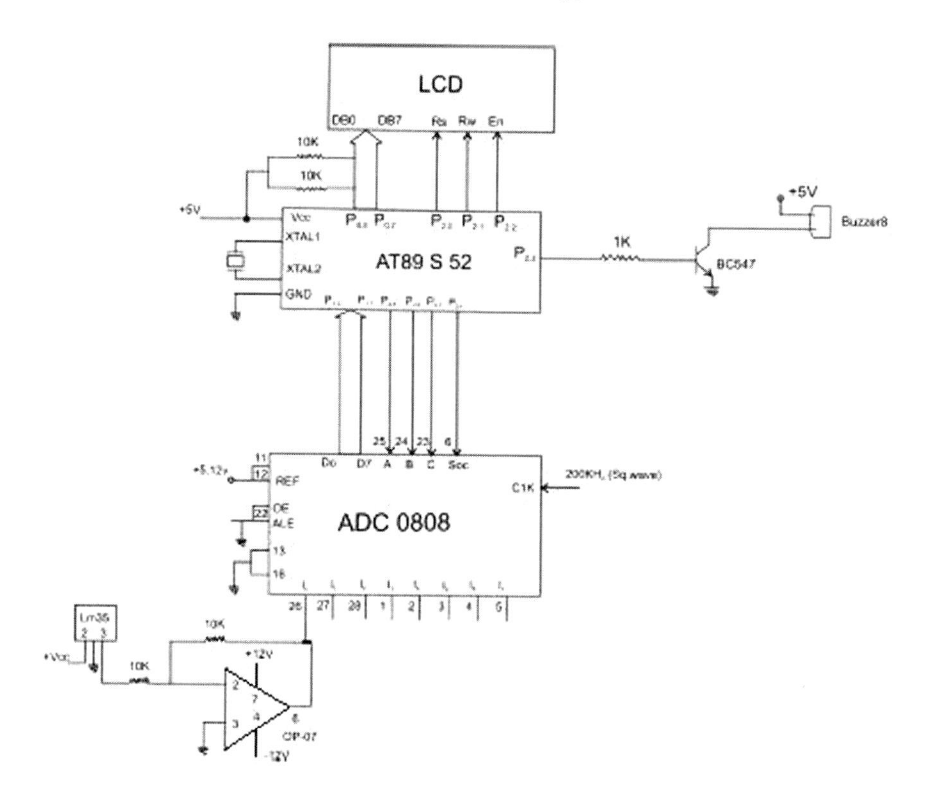

Figure 6.4 Circuit diagram of microcontroller based chimney sentinel unit

6.5 Application 5: Who's First Timer

This unit is ideal as a timer for quiz competitions. The person who first presses the key will be detected and his or her name will be displayed on LCD. The unit also sounds a buzzer to indicate that key has been pressed. The application is designed for eight users but can be modified easily to accommodate mode number of users.

Program Source Code

```
*****************************************************

#include <REG52.H>        /* special function register declarations */
#include <stdio.h>        /* for the intended 8051 derivative */

sbit RS = P2^0;
sbit RW = P2^1;
sbit EL = P2^2;
```

```
sbit R1 = P0^0;
sbit R2 = P0^1;

sbit R3 = P0^2;
sbit R4 = P0^3;

sbit C1 = P0^4;
sbit C2 = P0^5;
sbit C3 = P0^6;
sbit C4 = P0^7;

sbit buzz1= P3^0;
sbit buzz2= P3^1;
sbit buzz3= P3^2;
sbit buzz4= P3^3;

sbit buzz5= P3^4;
sbit buzz6= P3^5;
sbit buzz7= P3^6;
sbit buzz8= P3^7;

void delay(int);
void INIT(void);
void ENABLE(void);
void LINE(int);
int keyb(void);

void LINE(int i){
        if (i==1) {
        RS=0;
        RW=0;
        P1=0x80;
        ENABLE();
        RS=1;
        }
        else
        {
        RS=0;
        RW=0;
        P1=0xC0;
        ENABLE();
        RS=1;
        }
```

```
}

void delay(int k)
{
int i,j;
for (j=0;j<k+1;j++){
for (i=0;i<100;i++);
}}

void ENABLE(void)
{
EL=1;
delay(1);
EL=0;
delay(1);
}

void INIT(void)          // Initialization of the LCD by giving the proper
                            commands.
{
        RS=0;
        RW=0;
        EL=0;
        P1 = 0x38;    // 2 lines and 5*7 matrix LCD.
        ENABLE();
        ENABLE();
        ENABLE();
        ENABLE();
        P1 = 0x06;    //Shift cursor to left
        ENABLE();
        P1 = 0x0E;    // Display ON, Cursor Blinking
        ENABLE();
        P1 = 0x01;    // Clear display Screen
        ENABLE();
}

int keyb(void){                              //8user
int key=0;
C1=1;
C2=1;
C3=1;
C4=1;
```

```
R1=0;
R2=1;
R3=1;
R4=1;

if (C1==0)
{
key = 1;
buzz1=0;
}
if (C2==0)
{
key = 2;
buzz2=0;
}
if (C3==0)
{
key = 3;
buzz3=0;
}

if (C4==0)
{
key = 4;
buzz4=0;
}
R1=1;
R2=0;
R3=1;
R4=1;

if (C1==0)
{
key = 5;
buzz5=0;
}
if (C2==0)
{
key = 6;
buzz6=0;
}
if (C3==0)
```

```
{
key = 7;
buzz7=0;
}
if (C4==0)
{key = 8;
buzz8=0;
}
return(key);
}

void main(void) {
char array1[]="jivan";
char array2[]="kunal";
char array3[]="rupesh";
char array4[]="chari";
char array5[]="roy";
char array6[]="jesni";
char array7[]="vinod";
char array8[]="john";

char *p,key;
int j,i;

INIT();
LINE(1);
j=0;
while(1){
j=0;
j = keyb();

if(key==1){
for (i=0;i<8;i++){
 p=&array1;
P1= *p++;
ENABLE();

}}
if(key==2){
for (i=0;i<8;i++){
 p=&array2;
P1= *p++;
ENABLE();
```

```
}}
if(key==3){
for (i=0;i<8;i++){
p=&array3;
P1= *p++;
ENABLE();
}}
if(key==4){
for (i=0;i<8;i++){
 p=&array4;
P1= *p++;
ENABLE();
}}
if(key==5){
for (i=0;i<8;i++){
 p=&array5;
P1= *p++;
ENABLE();
}}
if(key==6){
for (i=0;i<8;i++){
 p=&array6;
P1= *p++;
ENABLE();
}}
if(key==7){
for (i=0;i<8;i++){
 p=&array7;
P1= *p++;
ENABLE();

}}
if(key==8){
for (i=0;i<8;i++){
 p=&array8;
P1= *p++;
ENABLE();
}}
delay(400);
}
}
*****************************************************
```

```
void delay(int k) // Delay function
{
int i,j;
for (j=0;j<k+1;j++)        // Delay loop
{
for (i=0;i<10000;i++);
}
}

void ENABLE(void) // Function to Enable the LCD
{
EL=1;                      // Give high to low pulse to enable pin of
                             the LCD
delay(1);
EL=0;
delay(1);
}

void INIT(void) // Function to Initialize the LCD
{
        RS=0;
        RW=0;
        EL=0;
        P1 = 0x38;        // LCD of 2 lines and 5x7 matrix
        ENABLE();
        ENABLE();
        ENABLE();
        ENABLE();
        P1 = 0x06;        // Shift cursor to right
        ENABLE();
        P1 = 0x0E;        // Display on, cursor blinking
        ENABLE();
        P1 = 0x01;
        ENABLE();
}

void main (void)
{
int count=0;
```

```
int k;
char *p;
char array[]="No of Cars="; /* Array to display the characters "No
                                  of Cars=" on the LCD */
INIT();

while(1)
{
count= updatecount(); /* Assign the returned new value to update the
                         count */
p=&array;                /* Points the next location on the LCD to
                         display the character*/
for (k=0;k<17;k++)
{
if (k==0)
LINE(1);
P1= *p++;              // Increment the pointer
ENABLE();
LINE(2);
unit =(count%10);
ten=(count/10)%10;     /* Display no of cars and the count on LCD
                       on respective decimal position. The character
                       to display should be in ASCII form.*/
hundred=(count/100)%10;
tenth=(count/1000);
P1= (tenth+0x30);        // Send the ASCII data to LCD
ENABLE();
P1= (hundred+0x30);
ENABLE();
P1= (ten+0x30);
ENABLE();
P1= (unit+0x30);

ENABLE();

}}
}
*****************************************************
```

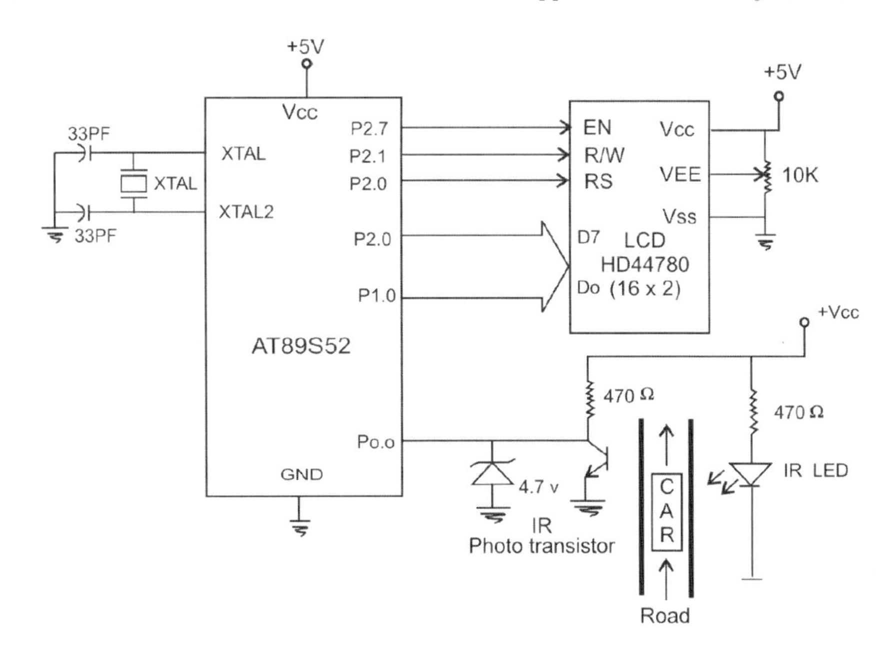

Figure 6.6 Circuit diagram of the system for counting cars

6.7 Application 7: Anonymous Voting

A useful device for taking the audience poll regarding a lecture or any other issue is described in this application. This device will be passed on to each and every member of the audience one by one. A message appears on the LCD as regards to whether the lecture is satisfactory or otherwise. Accordingly the user has to press key 1 or 2. The poll will be displayed as soon as key 3 is pressed.

Program Source Code

```
****************************************************

#include <REG52.H>          /* special function register declarations */
#include <stdio.h>          /* for the intended 8051 derivative */

sbit RS = P2^0;              // LCD control signals
sbit RW = P2^1;
sbit EL = P2^2;

sbit BU = P2^4;               // Buzzer as a output

sbit R1 = P0^0;              // Hex keypad connections
```

```
sbit R2 = P0^1;
sbit R3 = P0^2;
sbit R4 = P0^3;

sbit C1 = P0^4;
sbit C2 = P0^5;
sbit C3 = P0^6;
sbit C4 = P0^7;

void Delay(int);
void INIT(void);
void ENABLE(void);
void LINE(int);
int keyb(void);

void LINE(int i)          /* Function for the LCD display line selection
                          and giving the respective commands to LCD */
{
        if (i==1) {
        RS=0;
        RW=0;
        P1=0x80;
        ENABLE();
        RS=1;
        }
        else
        {
        RS=0;
        RW=0;
        P1=0xC0;
        ENABLE();
        RS=1;
        }
}

void delay(int k)                          // Delay function
{
int i,j;
for (j=0;j<k+1;j++){
for (i=0;i<100;i++);
}}
```

```
void ENABLE(void)       // Function to Enable the LCD. Give high
                           to low pulse on EL.
{
EL=1;
delay(1);
EL=0;
delay(1);
}

      void INIT(void)   /* Initialization of the LCD by sending the
                        commands sequentially */
{
      RS=0;
      RW=0;
      EL=0;
      P1 = 0x38;
      ENABLE();
      ENABLE();
      ENABLE();
      ENABLE();
      P1 = 0x06;
      ENABLE();
      P1 = 0x0E;
      ENABLE();
      P1 = 0x01;
      ENABLE();
}

int keyb(void)          // Keyboard scanning function. Checks which
                           key is pressed.
{
int key=0;

C1=1;
C2=1;
C3=1;
C4=1;

R1=0;
R2=1;
R3=1;
R4=1;
```

```
if (C1==0) key = 1;
if (C2==0) key = 2;
if (C3==0) key = 3;
if (C4==0) key = 4;

R1=1;
R2=0;
R3=1;
R4=1;

if (C1==0) key = 5;
if (C2==0) key = 6;
if (C3==0) key = 7;
if (C4==0) key = 8;

R1=1;
R2=1;
R3=0;
R4=1;

if (C1==0) key = 9;
if (C2==0) key = 10;
if (C3==0) key = 11;
if (C4==0) key = 12;

R1=1;
R2=1;
R3=1;
R4=0;

if (C1==0) key = 13;
if (C2==0) key = 14;
if (C3==0) key = 15;
if (C4==0) key = 16;
return(key);
}

void main(void)
{
char vote1[]="Lecture is satisfactory: Press 1"; /* Array for the
                                                   characters to display
                                                   on the LCD. */
char vote2[]="Lecture is not up to mark: Press 2";
```

```
char *p;                    // Pointer to point the next character in the
                              array
int j,i;
INIT();
LINE(1);
j=0;
while(1)
{            /* When key 1 is pressed matter raised is ok */
j=0;
j = keyb();
if (j==1){
p=&vote1;  // Memory address define to array and defined a pointer to
             point that data.
for(i=0;i<17;i++)
{
if(i==0)              // Display the data on the first line on the LCD
                        screen
LINE(1);
if(i==8)
LINE(2);    // Display the data on the second line on the LCD screen
P1= *p++;  // Pointer is incremented and the data pointed by pointer
             is sent on the port 1.
ENABLE();
Delay(200);
}
if (j==2){            /* If key 2 is pressed then the vote is for the matter
                        is not enough */
p=&vote2;
for(i=0;i<17;i++)
{
if(i==0)
LINE(1);
if(i==8)
LINE(2);    // Display the content on line 2
P1= *p++;  // Increment the pointer
ENABLE();
Delay(200);
}
}
}}}     //End of the main
```

****

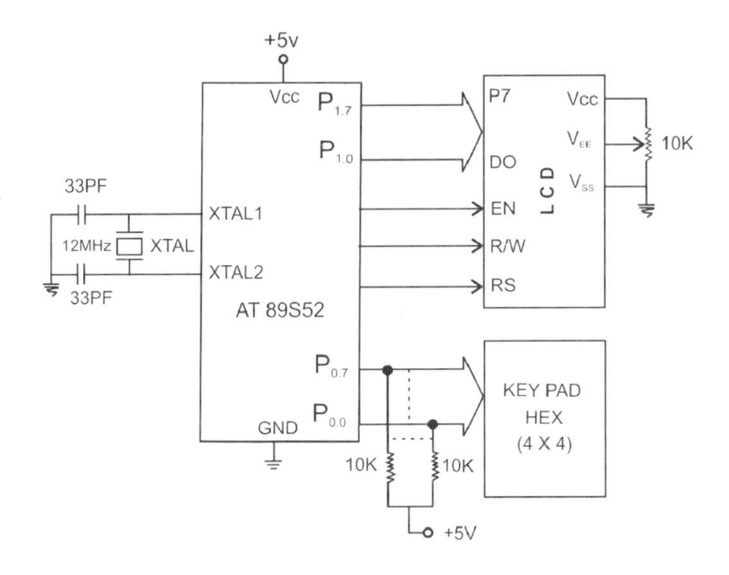

Figure 6.7 Circuit diagram of unit for anonymous voting

6.8 Energy Efficient Lighting Using Microcontroller

Sometimes the incandescent light bulb is referred to as a "heat bulb," because 90% of the electricity delivered to it comes out as heat. Precisely we have used this property to build a laboratory type PID application to illustrate the code in Chapter 6. However, the theme of the present chapter is energy efficient lighting using microcontrollers. With the growing concern about the energy crisis, almost all the industrialized countries are now coming out with new methods to save energy, especially electrical energy. It is estimated that the yearly usage of lighting for a single office room is 1000 h which leads to 20 KWh/m^2 for a light source which gives 20 W/m^2. With a microcontroller-based light switching or even dimming applications a 50% savings can be achieved. Recently, Freescale Semiconductor has come out with a new 8-bit HC908LB8 microcontroller that offers single-chip solution for energy-efficient lighting systems. However, the applications developed in this chapter are based on AT89S52 microcontroller.

Application 8: Optimize the Electric Power Consumption in the Corridors

In the long corridors of the hotels, unnecessary electricity is consumed on the lights. Ideally when no one is there the lights should be switched off. As the person passes from one end to the other the corresponding

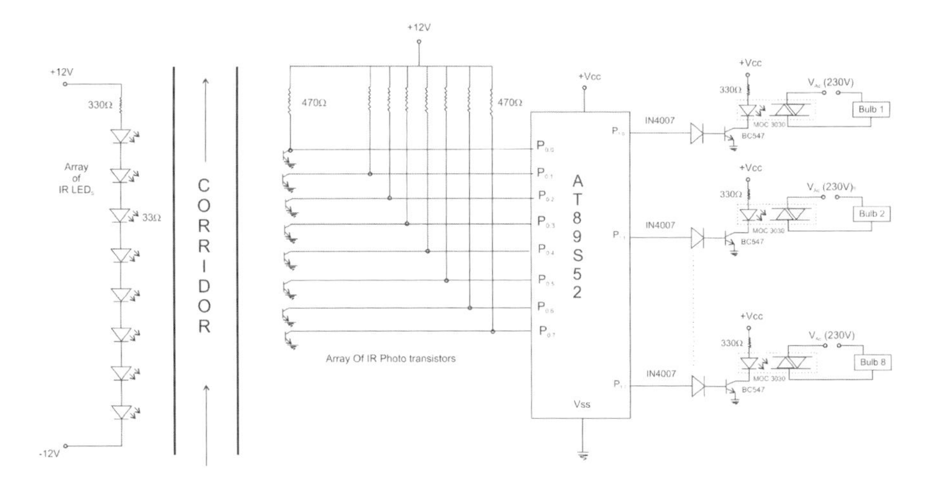

Figure 6.8 Circuit diagram of the system for auto switching of lights in the corridor

lights should be switched on so that he or she will be guided toward the corridor end. This problem is solved by using arrays of pairs of IR LEDs and phototransistor throughout the corridor placed at equal interval as shown in the block schematic. The array of IR LEDs are in a continuous emitting mode. The corresponding IR phototransistors are aligned at 45° for maximizing sensitivity. Value of the resistance (1 MΩ to 470Ω) connected to the collector of the phototransistor decides its sensitivity. When the IR link breaks due to the person passing the corridor, a low to high going transition is detected by the port 0 lines. The corresponding bulbs connected to port 1 are made ON and OFF so as to light the corridor as the person makes his way to the other end.

Program Source Code

```
*******************************************************

#include <REG52.H>    /* special function register declarations */
#include <stdio.h>
/* for the intended 8051 derivative   */

sbit sensor1= P0^0;       /* Output from the 8 sensors is connected to
                          the port pins of the port 0 */


sbit sensor2= P0^1;
sbit sensor3= P0^2;
sbit sensor4= P0^3;
```

```
sbit sensor5= P0^4;
sbit sensor6= P0^5;
sbit sensor7= P0^6;
sbit sensor8= P0^7;

/* Connecting the relays to on/off the power to the port 1 */
sbit RL1= P1^0;
sbit RL2= P1^1;
sbit RL3= P1^2;
sbit RL4= P1^3;
sbit RL5= P1^4;
sbit RL6= P1^5;
sbit RL7= P1^6;
sbit RL8= P1^7;

void main (void)            // Main function
{
P0=0x00;
P1=0x00;
while(1)
{
if(sensor1==1)              // Sense the first sensor

RL1=1;                      // If its output is high then switch the relay on

else if (sensor2==1) // otherwise check the second sensor.
{
RL1=0;      // Off relay 1 and on relay 2
RL2=1;
}

else if (sensor3==1)
{
RL3=1;
RL2=0;
}
else if (sensor4==1)
{
RL4=1;
RL3=0;
}
else if (sensor5==1)
```

```
{
RL5=1;
RL4=0;
}
else if (sensor6==1)
{
RL6=1;
RL5=0;
}
else if (sensor7==1)
{
RL7=1;
RL6=0;
}
else if (sensor8==1)
{
RL8=1;
RL7=0;
}
}

}
```

****

Chapter 7

Microcontroller-based Measurement and Control Applications

7.1 Application 1: Reading a PWM Waveform Using Microcontroller

There are many applications in which the microcontroller is used as a PWM generator for controlling purpose. For instance, the servo motor control or even the temperature controller can be driven on the basis of ON time modulation. However, in the following application, the PWM waveform generated by IC 555 timer is monitored by the microcontroller AT89S52 through port pin 2.4. The ON-time and OFF-time of the PWM are displayed on the LCD. The application may be used to interface resistive sensors like thermistor to display the temperature directly. The main advantage of the PWM-based measurement is the possibility of theoretically infinite resolution of the measurement.

Program Source Code

```
*****************************************************

#include <REG52.H>        /* special function register declarations */
                          /* for the intended 8051 derivative      */

sbit RS = P20;
sbit RW = P2^1;
sbit EN = P2^2;

sbit IN = P2^4;           /*input from 555 oscillator*/

void delay(int);
void INIT(void);
void ENABLE(void);
void LINE(int);
```


J.S. Parab et al. (eds.), Exploring C for Microcontrollers, 123–138.

```
int ONTIME(int);
int offtime(int);

void LINE(int i){
        if (i==1) {
        RS=0;
        RW=0;
        P1=0x80;
        ENABLE();
        RS=1;
        }
        else
        {
        RS=0;
        RW=0;
        P1=0xC0;
        ENABLE();
        RS=1;
        }
}

void delay()
{
int i,j;
for (j=0;j<10;j++){
for (i=0;i<100;i++);
}}

void ENABLE(void)
{
EN=1;
delay();
EN=0;
delay();
}

void INIT(void)    // Initialization of the LCD by giving the proper
                     commands
{
        RS=0;
        RW=0;
        EN=0;
```

```
        P1 = 0x38;    // 2 lines and 5*7 matrix LCD
        ENABLE();
        ENABLE();
        ENABLE();
        ENABLE();
        P1 = 0x06;    //Shift cursor to left
        ENABLE();
        P1 = 0x0E;    // Display ON, Cursor Blinking
        ENABLE();
        P1 = 0x01;    // Clear display screen
        ENABLE();
}

int ONTIME(int m){
if (m==1)                         // to measure On time
{
IN=1;                                  /*input to 8052 from 555 osc*/
TMOD=0x01;
TR0=0;
TF0=0;
TL0=0x00;
TH0=0x00;
while(IN);                      // check for rising edge
while(!IN);

TR0=1;
while(IN);                        // check for falling edge
TR0=0;
}
if(m==2)                        /* to measure offtime*/
{
IN=1;
TMOD=0x01;
TR0=0;
TF0=0;
TL0=0x00;
TH0=0x00;

while(!IN);          // check for falling edge
while(IN);

TR0=1;
```

```
while(!IN);            // check for rising edge
TR0=0;

}
return((TH0*256)+TL0);

}
void main (void){
int unit, tens, hundred, thousand, tenthou, x;

char code dis3[]="Fre in HZ=";
char *p;
int i,j,k,freq,l;
while(1) {
for (x=1;x<=3;x++){
if( x==1)
{
i = ONTIME(x);
k=i;
}
if( x==2)
{
i = offtime(x);
l=i;
}
if (x==3)
{
freq=k+l;
i=1/freq ;                              //convert time to frequency
}

unit = (i%10);
tens = (i/10)%10;
hundred = (i/100)%10;
thousand = (i/1000)%10;
tenthou = (i/10000)%10;
unit = unit + 0x30;
tens = tens + 0x30;
hundred = hundred + 0x30;
thousand = thousand + 0x30;
tenthou = tenthou + 0x30;
```

```
INIT();
LINE(1);

p=&dis3;                        // display freq on LCD

for (j=0;j<8;j++){
        P1= * p++;
        ENABLE();
        }

LINE(2);
P1=tenthou;
ENABLE();
P1=thousand;
ENABLE();
P1=hundred;
ENABLE();
P1=tens;
ENABLE();
P1=unit;
ENABLE();
}
}}
*****************************************************
```

7.2 Single Set-point On/Off Controller

An on/off controller is the simplest form of temperature control device. In the following application, the process consists of a simple heating unit based on Neon bulb. The sensor used is LM35, the output of which is conditioned by using OP07. The digitization is done by using ADC7135. The end of conversion is sensed by using the port pin 3.4 of AT89S52. The data is taken in using port P0. The set-point is kept at 40, but the same can be varied based on the application needs. The LCD displays the set-point as well as current temperature. The controlling is done by using a relay driven by port line P2.3 through transistor 2N2222.

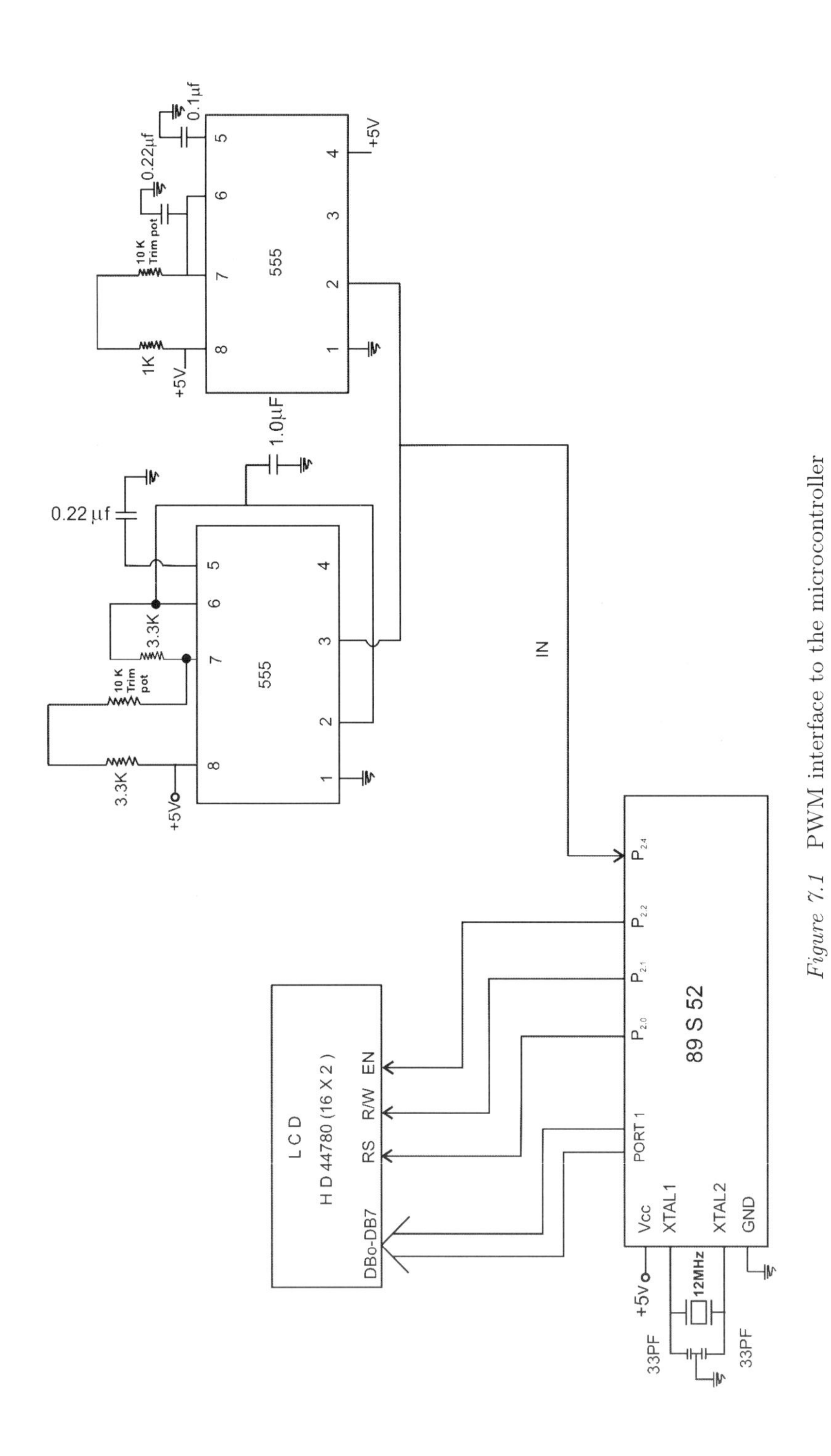

Figure 7.1 PWM interface to the microcontroller

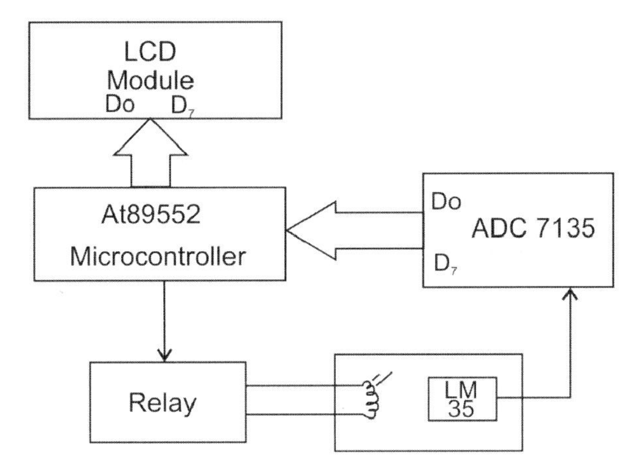

Figure 7.2 Block schematic of microcontroller based on/off controller

Program Source Code

```
*************************************************
#include<REG52.H>
void busy_count(void);

void LINE(int a);
void ENABLE(void);
void init(void);
//void init_lcd(void);
sbit clk = P3^5;
sbit BUSY=P3^4;
sbit RS = P2^0;
sbit RW = P2^1;
sbit EL = P2^2;
sbit RL = P2^3;
int count_u, count_t;

void main()
{
 TMOD=0x50;
 BUSY=1;
 init();
 busy_count();   /*start timer in function*/
}
```

```
void busy_count(void)
{
int count, count1;
 TR0=0;
 TH0=0x00;
 TL0=0x00;
 count=0x00;
 while(BUSY);
 while(!BUSY);
 while(BUSY)
   {
   TR0=1;        /* start counter*/
        }
 while(!BUSY)
        {
        TR0=0;          /*when busy goes low stop counter*/
        }
 count=((TH0*256)+TL0);
count1=count-10000;
count_u=count%10;
count_t=(count/10)%10;
RS=1;
LINE(1);
if (count >40)        //if temperature is greater than 40 than turn off
                        the relay
{
RL=1;
}
else
{
RL=0;
}
P1=(count_t+0x30);
ENABLE();
P1=(count_u+0x30);
ENABLE();
}

void init(void){
        RS=0;
        RW=0;
        EL=0;
```

```
        P1 = 0x38;
        ENABLE();
        ENABLE();
        ENABLE();
        ENABLE();
        P1 = 0x06;
        ENABLE();
        P1 = 0x0E;
        ENABLE();
        P1 = 0x01;
        ENABLE();

}

void LINE(int i){
        if (i==1) {
        RS=0;
        RW=0;
        P1=0x80;
        ENABLE ();
        RS=1;
        }
        else
        {
        RS=0;
        RW=0;
        P1=0xC0;
        ENABLE();
        RS=1;
        }
}
void ENABLE(void)
{
EL=0;
//delay(1);
EL=1;
//delay(1);
EL=0;
}
****************************************************
```

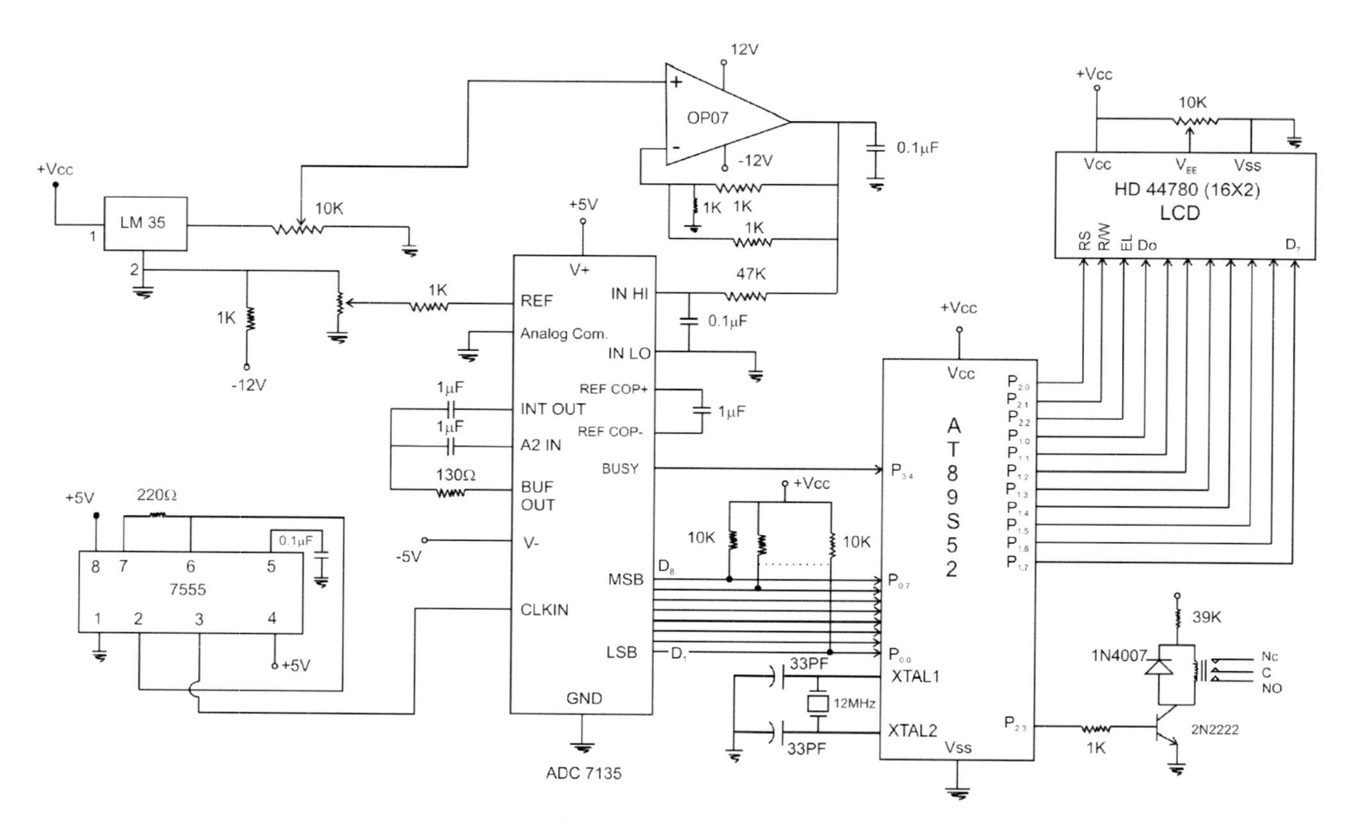

Figure 7.3 Circuit diagram of microcontroller-based on/off controller

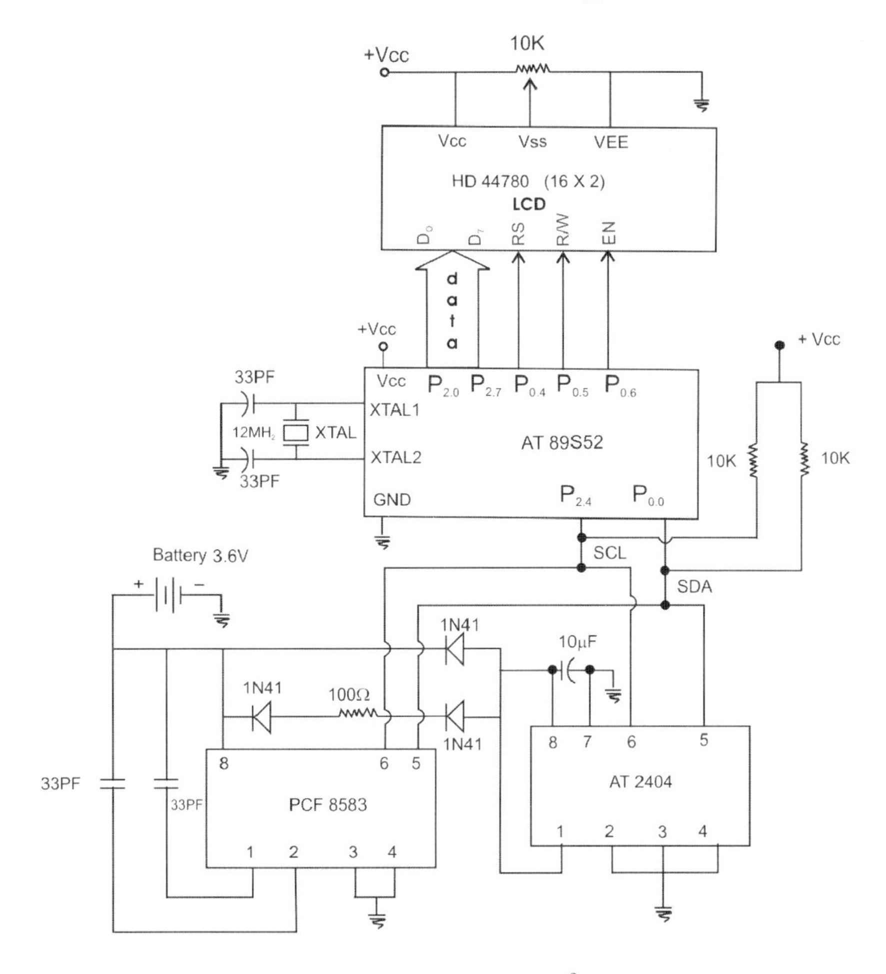

Figure 7.4 Circuit diagram for I^{2}C interface

7.3 Application 3: I^{2}C Interface with Serial EPROM

In the context of microcontroller-based process control applications, many embedded designers are experiencing the inadequacy of the on-chip memory. The simplest alternative is to interface an external EPROM. I^{2}C-based serial peripherals offers maximum circuit efficiency with the interface simplicity. The prime advantage of this interface is that only two lines (clock and data) are required for full duplexed communication between multiple devices. The operating range of the interface is typically 100 KHz to 400 KHz. I^{2}C protocol assigns a unique address to every peripheral. The peripheral can act as a receiver and/or transmitter depending on its functionality.

The application developed here illustrates the I²C-enabled code development for the PCF8583 Real-Time Clock and AT2404 Serial EPROM. In many data logging applications the data needs to be collected with the time stamp. The RTC PCF8583 will facilitate the time stamping operation while the acquired data can be stored in the serial EPROM AT2404.

Program Source Code

```
*************************************************

#include <reg52.H>

sbit sda=P0^0;
sbit scl=P0^1;
sbit led=P1^0;

sbit RS=P0^4;
sbit RW=P0^5;
sbit EN=P0^6;

int x;
unsigned char dat;

int ack(void);
void start(void);
void stop(void);
void delay(void);
void enable(void);
void add(int);
void ini(void);
void reset(void);
unsigned char read(void);
void lcd(unsigned char);

void start(void)
{
sda=1;
delay();
scl=1;
delay();
sda=0;
delay();
}
```

```
void stop(void)
{
sda=0;
delay();
scl=1;
delay();
sda=1;
delay();
}

void delay(void)
{
int i;
for(i=0;i<30000;i++);
for(i=0;i<200;i++);
}

void add(d)
{
int i;
for(i=0;i<8;i++)
{

scl=0;
delay();
sda=d>>7;
scl=1;
delay();
d=d<<1;
}
scl=0;
delay();

}

int ack(void)
{
scl=0;
delay();
sda=1;
delay();
```

```
scl=1;
delay();
return sda;
}
void reset(void)
{
int i;
for(i=0;i<9;i++)
{
scl=0;
delay();
scl=1;
delay();
if(sda==1)
break;

}
}

unsigned char read(void)
{
unsigned char var=0;
int i;
for(i=0;i<8;i++)
{

scl=0;
delay();

scl=1;
var=var|sda;
delay();
var=var<<1;
}
scl=0;
delay();
return var;
}

void ini(void)
{
RW=0;
RS=0;
```

```
P2=0x38;
enable();

enable();
enable();
enable();
P2=0x06;
enable();
P2=0x0E;
enable();
P2=0x01;
enable();
RS=1;

}
void lcd(unsigned char dat)
{
int hun, ten, unit;
hun=dat/100;
ten=((dat/10)%10);
unit=dat%10;

hun=(hun+0x30);
P2=hun;
enable();
delay();
ten=ten+0x30;
P2=ten;
enable();
delay();
unit=unit+0x30;
P2=unit;
enable();
delay();

}

void enable(void)
{
EN=1;
delay();
EN=0;
delay();
}
```

```
void main(void)
{
ini();

while(1)
{
P1=0x00;
reset();
start();

add(0xA0);
led=1;
x=ack();
led = x;
delay();
delay();
led=1;
if (x==0)
        {
        add(0x05);
        x=ack();
        led = x;
        delay();
        delay();
        }
led=1;


start();
add(0xA1);

dat=read();
lcd(dat);
scl=1;
sda=1;

stop();

}
}
****************************************************
```

Chapter 8

Securing Your Embedded System Application

8.1 Security Challenges in Embedded Systems

The applications of embedded systems are growing in the areas such as cell phones, cars, spacecraft, biomedical sensitive, and defense. The main concern for deployment in these application sectors is security. Security has been the subject of intensive research in the context of general-purpose computing and communications systems. In computing and networking products there are established devices and algorithmic methodologies for ensuring the security of the code. The security concerns about the embedded systems are well covered in many references [38; 39; 40; 41; 42; 43]. Security is often misconstrued by embedded system designers as the addition of features such as specific cryptographic algorithms and security protocols to the system. In reality, it is a new dimension that designers should consider throughout the design process, along with other metrics such as cost, performance, and power. The challenges unique to embedded systems require new approaches to security covering all aspects of embedded system design from architecture to implementation. Security processing, which refers to the computations that must be performed in a system for the purpose of security, can easily overwhelm the computational capabilities of processors in both low- and high-end embedded systems [38]. This challenge, which we refer to as the "security processing gap," is compounded by increases in the amounts of data manipulated and the data rates that need to be achieved [38]. Equally daunting is the "battery gap" in battery-powered embedded systems, which is caused by the disparity between rapidly increasing energy requirements for secure operation and slow improvements in battery technology. The final challenge is the "assurance gap," which relates to the gap between functional security measures (e.g., security services, protocols, and their constituent cryptographic algorithms) and actual secure implementations.

J.S. Parab et al. (eds.), Exploring C for Microcontrollers, 139–150.

The case studies developed in this chapter specifically focus on the security aspects of the embedded systems. The first case study offers the security by checking the user's login ID and password just like a typical interface presented by multi-user OS. The second case study allows the user to let in provided he or she enters the password in a given time. While the third application illustrates the methodology of writing the routines in the same manner as that of computer BIOS, which has inherent code security.

8.2 Application 1: Authentication for Your Embedded System Application

The setup shown in Figure 8.1 is used for this application. The program prompts the user to enter his user ID by displaying a message "LOGIN" on the LCD. The user is expected to enter his ID using the PC keyboard. Then the program displays a message "password" on the LCD to prompt the user to enter the password. After entering the password, the program will check whether the entered information is correct or not.

The system can also be used for cabinet lock-up or for any keyboard-based security implementation.

Program Source Code

```
*************************************************

#include <REG52.H>          /* special function register declarations */
                              / * for the intended 8051 derivative */

#include <stdio.h>     /* prototype declarations for I/O functions */

sbit RS = P2^0;
sbit RW = P2^1;
sbit EL = P2^2;

void INITt ();
void delay(int);
void INIT(void);
void ENABLE(void);
void LINE(int);
char array[]="LOGIN:";
char pass[]="EMBEDED";
char Access[]="CORRECT";
char denied[]="WRONG";
```

```
void INITt()          // function to Initialize the timer
{
        TMOD=0x20;
        TH1=0xFD;
        SCON=0x50;
        TR1=1;


}
void LINE (int i)     // Enable the line function
        {
        if (i==1)
                  {
                  RS=0;
                  RW=0;
                  P1=0x80;
                  ENABLE();
                  RS=1;
                  }
        else
                  {
                  RS=0;
                  RW=0;
                  P1=0xC0;
                  ENABLE();
                  RS=1;
                  }
        }

void delay(int k)           // invoking delay function
        {
        int i,j;
        for (j=0;j<k+1;j++)
                  {
        for (i=0;i<10000;i++);
                  }
        }

void ENABLE(void)       // invoking the enable function
        {
        EL=1;
```

```
        delay(1);
        EL=0;
        delay(1);
        }

void INIT(void)            /*initialization of LCD display*/
        {
        RS=0;
        RW=0;
        EL=0;
        P1 = 0x38;
        ENABLE();
        ENABLE();
        ENABLE();
        ENABLE();
        P1 = 0x06;
        ENABLE();
        P1 = 0x0E;
        ENABLE();
        P1 = 0x01;
        ENABLE();
        }

void main (void)           // main function
        {
        char b,var;
        char *p;
        int i,j,k;
while (1)
        {
        INIT();
        LINE (1);
        for (b=0;b<8;b++)
                {
                p=&array;
        P1= *p++;      /* increment the pointer and point the
                                         data from the array */

        ENABLE ();
                }
```

```
{
INITt();
RI=0;
var=SBUF;

while(!RI);
RI=0;
while(1)
{
for (i=0;i<8;i++)
{

if((pass[i])==(var[i]))
for(j=0;j<8;j++)
{
LINE(1);
p=&Access;
P1= *p++;
ENABLE();
}
else
{
LINE(1);
p=&denied;
P1= *p++;
ENABLE();
}

RS=0;
P1=0x01;
ENABLE();
RS=1;
}
}
}}}
```

****

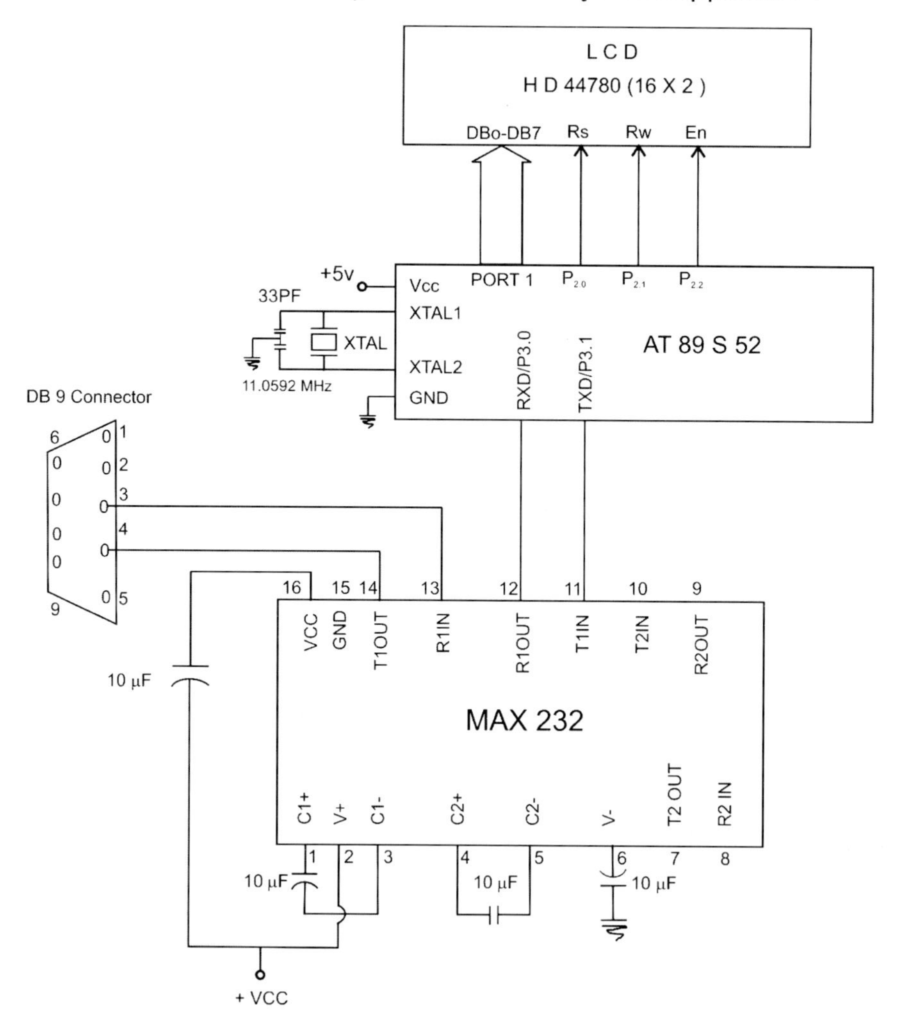

Figure 8.1 Setup for security authentication of embedded system

8.3 Application 2: Timeout Waiting for Input Data

In this application, the security is implemented by allowing the user to enter 10 predetermined characters (password or secret word) in a predetermined time period. Thus this application ensures a tight security implementation as compared to the previous as the time clause is important.

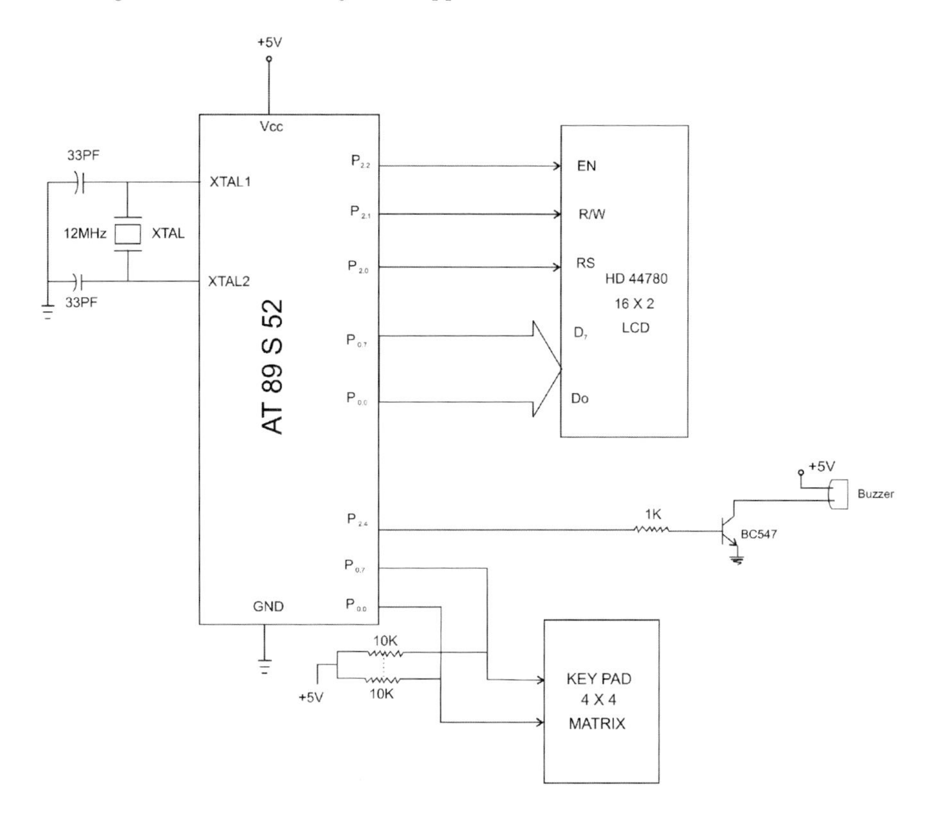

Figure 8.2 Schematic for the setup for timeout waiting

It can also be used for normal data logging applications. While inputting the data, if the data does not arrive after some allowed time, the wait is abandoned and some other action is taken. The request can be retransmitted or some error message is sent to the user.

The program implemented here waits for the user to type 10 characters. The timeout for the first character is 10 seconds, and then 2 seconds for each character thereafter. If the user does not respond within this time, a timeout message is displayed on LCD.

Program Source Code

```
*************************************************************

#include <REG52.H>      /* special function register declarations */
#include <stdio.h>
/* for the intended 8051 derivative          */
```

```
sbit RS = P2^0;      // LCD control signals
sbit RW = P2^1;
sbit EL = P2^2;
sbit BU = P2^4;      // Buzzer

sbit R1 = P0^0;      // Keypad connection 4x4 matrix keyboard
sbit R2 = P0^1;
sbit R3 = P0^2;
sbit R4 = P0^3;

sbit C1 = P0^4;
sbit C2 = P0^5;
sbit C3 = P0^6;
sbit C4 = P0^7;

void delay(int);           /* Stop Exection with Serial Intr. */
void INIT(void);
void ENABLE(void);
void LINE(int);
int keyb(void);
char getascii(int);    // ASCII convertion to send data on LCD

char getascii(int k)    // Invoking the ASCII conversion function
{

        int ascii;
        if(k<10)
        ascii=k+0x30;     // converts the data in to ASCII format
        if (k==10)
        ascii=0x30;
        return (ascii);     // Returns the ASCII value to the main
                              routine
}


void LINE(int i)      // function for selecting the line to display
                        the data on LCD
{
        if (i==1)
        {
        RS=0;
```

```
        RW=0;
        P1=0x80;
        ENABLE();
        RS=1;
        }
        else
        {
        RS=0;
        RW=0;
        P1=0xC0;
        ENABLE();
        RS=1;
        }
}

void delay(int k)           // Delay function
{
int i,j;
for (j=0;j<k+1;j++)
{
for (i=0;i<100;i++);
}}

void ENABLE(void)        // Invoking the function to enable the LCD
{
EL=1;
delay(1);
EL=0;
delay(1);
}

void INIT(void)        // Sequential commands for the LCD
                          initialization
{
        RS=0;
        RW=0;
        EL=0;
        P1 = 0x38;
        ENABLE();
        ENABLE();
        ENABLE();
        ENABLE();
```

```
        P1 = 0x06;
        ENABLE();
        P1 = 0x0E;
        ENABLE();
        P1 = 0x01;
        ENABLE();
}

int keyb(void)          // Invoking the Keyboard scanning function
{
int key=0;

C1=1;
C2=1;
C3=1;
C4=1;

R1=0;
R2=1;
R3=1;
R4=1;

if (C1==0) key = 1;
if (C2==0) key = 2;
if (C3==0) key = 3;
if (C4==0) key = 4;

R1=1;
R2=0;
R3=1;
R4=1;

if (C1==0) key = 5;
if (C2==0) key = 6;
if (C3==0) key = 7;
if (C4==0) key = 8;

R1=1;
R2=1;
R3=0;
R4=1;
```

```
if (C1==0) key = 9;
if (C2==0) key = 10;
if (C3==0) key = 11;
if (C4==0) key = 12;

R1=1;
R2=1;
R3=1;
R4=0;

if (C1==0) key = 13;
if (C2==0) key = 14;
if (C3==0) key = 15;
if (C4==0) key = 16;
return(key);
}

void main(void)                    // Main function
{
int j;
char array[]="Enter 10 characters";
char time out[]="Request time out";
INIT();
char *p;
int k;
while(1){
p=&array;
for (k=0;k<17;k++)
{
if (k==0) LINE(1);
if (k==8) LINE(2);
P1= *p++;
ENABLE();
j = keyb();
Delay(10000);            //10 sec delay
If(j=0)              //no key pressed

{
p=&time out;
for (k=0;k<17;k++)
{
```

```
if (k==0) LINE(1);
if (k==8) LINE(2);
P1= *p++;
ENABLE();
}
elseif{
for (i=0;i<9;i++)          //for remaining 9 characters delay is 2 sec
{
P1=getascii(j);
delay(2000);               // 2 sec
j = keyb();
if(j=0)
{ p=&time out;             // if no character is received within 2 sec
                 // then flash "request time out on LCD screen
for (k=0;k<17;k++)
{
if (k==0) LINE(1);
if (k==8) LINE(2);
P1= *p++;
}
ENABLE();

}
}}
*************************************************************
```

References

1. http://en.wikipedia.org/wiki/Microcontroller from Wikipedia, the free encyclopedia
2. http://www.webopedia.com/TERM/M/microcontroller.html
3. http://www.definethat.com/define/1727.htm
4. http://www.pcmag.com/encyclopedia_term/0,2542,t=microcontroller&i=46924,00.asp
5. http://foldoc.org/?microcontroller
6. Report on World Microcontrollers Market by Frost & Sullivan research service, published on 30 December 2005
7. http://www.intel4004.com/busicom.htm A testimonial from Federico Faggin, its designer, on the first microprocessor's thirtieth birthday
8. http://www3.sk.sympatico.ca/jbayko/cpu1.html Section One: Before the Great Dark Cloud
9. http://www.cpushack.net/Historyofthe8051.html The Unofficial History of 8051 by Jan Waclawek (wek at efton.sk) edited by John Culver
10. World Microcontrollers Market F591-26, Agency/Source: Frost & Sullivan http://www.newswiretoday.com/news/7122/)
11. Applications for Efficiency: The Green Story, www.freescale.com
12. The 8-bit microcontroller-A hit product that supports digital still cameras (DSCs) from behind the scenes 8-bit microcontrollers and the important role they play inside digital cameras, Vol. 24 (14 September 2004) http://www.necel.com/en/channel/vol_0024/vol_0024_2.html#chapter1)

13. http://www.bourneresearch.com/: Bourne Research is a trusted source of market intelligence, with a specialized focus on MEMS (MicroElectroMechanical Systems), Nanotechnology, and the convergence of both.

14. SIA Raises Chip Sales Forecast, Written by Steven Waller, Wednesday, 07 June 2006 http://e-and-f.com/index.php?option=com_content&task=view&id=40&Itemid=2

15. http://www.cs.ucr.edu/content/esd/slide_index.html

16. http://www.industrialnewsroom.com/fullstory/26253 Cygnal Releases World's Highest Performance 8051 Microcontroller

17. http://www.xemics.com/ or http://www.semtech.com/products/wireless&sensing/trans/xemic.jsp XE8000 Application Specific Microcontrollers series

18. http://www.eeproductcenter.com/micro/brief/showArticle.jhtml?articleID=196700958 Microcontrollers integrate power-saving technology, Gina Roos, eeProductCenter, 12/19/2006

19. http://www.microcontroller.com Microcontrollers and DSPs – Will They Converge? by Bill Giovino

20. http://www.edn.com/index.asp?layout=article&stt=000&articleid=CA333669&pubdate=11%2F13%2F2003: Microcontrollers gain DSP attributes in "hybrid" architecture By Graham Prophet – EDN Europe, 11/13/2003

21. http://www.omimo.be/magazine/99q2/Hitachi.pdf The RISC Challenge in DSP Processing by Dr. Manfred Schlett

22. http://www.mcjournal.com/articles/arc101/arc101.htm Real-Time Debugging Highly Integrated Embedded Wireless Devices, by David Ruimy Gonzales and Brian Branson

23. http://www.keil.com/dd/chip/3469.htm Silicon Laboratories, Inc. C8051F120

24. www.futurlec.com/News/Dallas/InternetIC.html New Microcontroller is web-enabled Dallas Semiconductor Re-Engineers, Its Microcontrollers for Network Computing

25. http://www.aldec.com/products/ipcores/ IP cores by Aldec Inc.

26. http://blogs.zdnet.com/emergingtech/wp-mobile.php?p=227&more=1 Emerging Technology Trends

27. http://www.embedded.com/showArticle.jhtml?articleID=194300451 Tutorial: Improving the transient immunity of your microcontroller-based embedded design – Part 1 Defining the problem: A step by step tutorial on EMI, ESD, and EFT problems in embedded designs and a range of possible solutions. By Ross Carlton, Freescale Semiconductor, Inc

28. http://www.keil.com/support/man_c51.htm On Line Manual of Keil for C51 products

29. "Choosing a Microcontroller for Embedded System Applications", Mel Tsai http://www.egr.msu.edu/classes/ece482/Reports/appnotes/98spr/tsaimelv/appnote.html

30. Application Note: Choosing a Microcontroller for Embedded Systems Applications Mel Tsai http://www.mtsai.net/documents/appnote/appnote.html

31. http://www.airborn.com.au/8051/2wio8051.html 2 Wire Input/Output for 8051 type CPU's

32. www.8052.com/users/garypeek/ : "I/O EXPANSION ROUTINES FOR 8051 FAMILY MICROCONTROLLERS; WRITTEN BY GARY"

33. On the verge: LEDs ready to challenge incumbent light sources in the street lighting market: Tim Whitaker http://ledsmagazine.com/articles/features/3/10/4/1: LED's magazine

34. Microcontrollers provide connectivity of HB-LED lighting products: Dugald Campbell http://www.ledsmagazine.com/articles/features/2/11/1/1 LEDs Magazine

35. http://www.alldatasheet.co.kr/datasheet-pdf/pdf_kor/STMICROELECTRONICS/ULN2003.html Data sheet of ULN2003

36. www.aurel32.net/elec/pcf8583.pdf Data-sheet of PCFR8583

37. http://www.atmel.com/dyn/products/product_card.asp?part_id=2806 Data-sheet of AT24C04

38. Security in embedded systems: Design challenges, Srivaths Ravi, Anand Raghunathan, Paul Kocher, Sunil Hattangady, ACM Transactions on Embedded Computing Systems (TECS) archive, Volume 3, Issue 3 (August 2004) table of contents, pp. 461–491, 2004, ISSN:1539-9087

39. Ross J. Anderson, Markus G. Kuhn, Low-Cost Attacks on Tamper Resistant Devices, Proceedings of the 5th International Workshop on Security Protocols, p.125–136, April 07–09, 1997

40. W. A. Arbaugh, D. J. Farber, J. M. Smith, A secure and reliable bootstrap architecture, Proceedings of the 1997 IEEE Symposium on Security and Privacy, p.65, May 04–07, 1997

41. Matt Blaze, A cryptographic file system for UNIX, Proceedings of the 1st ACM Conference on Computer and Communications Security, pp. 9–16, 3–5 November 1993, Fairfax, Virginia, USA

42. D. Boneh, R. DeMillo, and R. Lipton, 2001. On the importance of eliminating errors in cryptographic computations. Cryptology 14(2): 101–119

43. Jerome Burke, John McDonald, Todd Austin, Architectural support for fast symmetric-key cryptography, Proceedings of the 9th International Conference on Architectural Support for Programming Languages and Operating Systems, pp.178–189, November 2000, Cambridge, Massachusetts, USA

Index

H

I

J

K

L

M

O

P

R

S